THE

ATTAC and DEFENCE

OF

FORTIFIED PLACES.

Also from Flower-de-Luce Books

The Practical Surveyor, the 1725 first Edition
by Samuel Wyld
A Treatise of Mathematical Instruments, the 1775 third Edition
by John Robertson

THE

ATTAC and DEFENCE
OF
FORTIFIED PLACES.

By JOHN MULLER,
Professor of Artillery and Fortification.

With NOTES
By DAVID MANTHEY.

A REPRINT of the
SECOND EDITION from 1757.

Flower-de-Luce Books
The Invisible College Press
Arlington, Virginia

The Attac and Defence of Fortified Places

by John Muller

Original Second Edition published in London, England, 1757

Flower-de-Luce Edition
Editing and Notes by David Manthey

ISBN: 1-931468-18-4

First Printing

Flower-de-Luce Books
Published by
The Invisible College Press, LLC
P.O.Box 209
Woodbridge, Virginia 22194-0209
http://www.invispress.com

Please send questions and comments to
editor@invispress.com

This reprint is dedicated to the

New York Provincial Soldiers
who besieged Fort Niagara in 1759

and to the

Living-history reenactors who portray
Schuyler's Company of New York Provincials

THE ATTAC and DEFENCE OF FORTIFIED PLACES.

In three PARTS.

CONTAINING,

I. Preparations for, and Operations of an Attac, from the Beginning to the End, in a regular and eafy Manner.

II. Preparations for, and Defence of every Part of a Fortification, with all neceffaries for a good Defence.

III. A Treatife of MINES, explaining the Manner of making and loading them, with Tables of the proper Charges and Dimenfions of the Cubic Boxes for loading from 50 to 640 Pounds of Powder for their Charges, deduced from a new Theory.

For the Ufe of the Royal Academy of Artillery at *Woolwich*, and all concerned in the Art of War, by Land or Sea.

Illuftrated with Twenty-five large Copper-plates.

By JOHN MULLER,
Profeffor of Artillery and Fortification.

The SECOND EDITION,
Corrected and very much Enlarged, with new Tables, *&c.*

LONDON:
Printed for J. MILLAN, near *Charing Crofs*, 1757.
[Price Six Shillings.]

To his ROYAL HIGHNESS,

William Duke of *Cumberland*,

Captain-General and Commander in Chief of all his Majeſty's Forces, *&c.* *&c.* *&c.*

S I R,

THE favourable Attention and generous Protection with which your Royal Highneſs has conſtantly honoured thoſe who have uſed their beſt Endeavours to promote the Knowledge of any Branch of the Art of War, encourages me to preſent this Work to your Royal Highneſs, whoſe Genius and Application has made you the moſt competent Judge, as your high

high Birth and Command, claim the Patronage of ſuch Performances.

My Intention is to render the Knowledge of this important Part of the Art of War intelligible.

And that it may meet with your Royal Higneſs's favourable Approbation and Acceptance, is the higheſt Ambition of,

Your ROYAL HIGHNESS'S

Moſt Devoted

and moſt Humble Servant,

JOHN MULLER.

Royal Academy of Artillery at *Woolwich*, *July* 20th, 1757.

PREFACE.

IF, in the Art of War Courage alone was ſufficient, a Treatiſe on this Subject in *England* would be ſuperfluous; who from the loweſt to the higheſt Subject are celebrated for this Virtue; but as it oft happens, that Enemies that dare not appear in the open Field, may retire into fortified Places, and there by Art and Strategems deſtroy thoſe Troops which before they durſt not Face. For which Reaſon this Work was compiled, to furniſh Means to ſuch brave Men to avoid engaging on diſadvantageous Terms, and to excite in them a Deſire to apply their natural Genius to the Knowledge of all the different Occurrences which happen in their Station during the Courſe of a War, how to defend themſelves, or to attac an Enemy with Advantage, and to ſave Lives ſo much as poſſible, in conducting the Works of a Siege with Skill and Prudence. There is no Part of the Art of War that requires more Capacity, Knowledge, and Judgment. The Succeſs of a Battle is oft owing to Chance; a firſt Firing doing more Execution on

one

one ſide than the other, an advantageous Situation, a Party flanking the Army unperceived; ſome extraordinary Action performed by ſome Men or Officer; a ſudden Wind or heavy Rain in the Face of an Army; a pannick, Terror, or in ſhort any unforeſeen Accident, may contribute to the Gain or Loſs of a Battle, and that oft on the Side who had leaſt Reaſon to expect it.

Hiſtory, ancient and modern, furniſh ſo many Examples of this Kind, that it is needleſs to enumerate them. But when required to attac an Enemy in a Place ſtrongly fortified, or when the Fortune of War obliges a General to defend himſelf in ſuch a Place, againſt a numerous Army well provided, who find Means to approach the Works under Cover, beat them down with their Cannon, or blow them up with Mines, and ruin all their Defences; it is then that Courage and Fortune are obliged to ſubmit to Knowledge and Experience. There have been many brave Officers, provided with a good Garriſon, and every thing requiſite to make a brave Defence, who, thro' want of a competent Knowledge, were not able to preſerve a fortified Place to their Prince or Nation, or to take one; when, on the contrary, Men endued with leſs Bravery and more Knowledge, have ſucceeded in their Undertakings. Courage is undoubtedly a neceſſary Ingredient in the Compoſition of a Soldier, therefore every one ſhould feel his Pulſe well before he enters into this Way of Life, it being a Qualification not to be acquired. Nor is Experience alone ſufficient to form a good Commander; for the Situations and Conſtructions of Places are ſo different, and the Time and Occurrences ſo various, that it is impoſſible an Officer ſhould, with the bare

Experience

Experience of a few Sieges, be in a Condition to act with all the Conduct and Prudence neceſſary at any other Time, when perhaps every Part ſhould be done quite differently from what he might have ſeen before.

Let us ſuppoſe a General endued with great Bravery, but unacquainted with the Art of attacking and defending fortified Places, enters an Enemy's Country, and after having committed ſome Ravages meets the Army of his Adverſary and defeats it; which however is an Advantage not ſo eaſily gained, ſince, if his Enemy is a wary and prudent Commander, but inferior in Strength, he may poſſibly avoid an Engagement; but what Benefit will he reap from his Victory, if his Adverſary retires into ſome ſtrong Place, which he knows not how to attac and carry without the Loſs of the beſt Part of his Army? Should he advance farther into the Enemy's Country, and leave ſtrong Places behind him, he will run the Riſque of being ſurrounded on every ſide, or having his Convoys of Ammunition and Proviſion cut off, whereby his Army will be reduced thro' Famine to retire; and that perhaps with Shame and the Loſs of a conſiderable number of Men, to no Purpoſe, notwithſtanding his having been victorious at firſt.

It may be ſaid, that the Engineers will ſupply his want of Knowledge, by informing him which are the ſtrongeſt and the weakeſt Parts of the Place, how and in what Manner it is to be attacked, how the Approaches are to be made and carried on: This is certainly their Duty, and they ought undoubtedly to acquit themſelves in the beſt manner they are capable; but how oft happens it that the Engineers are as deficient as the General? how will

will he then proceed, or how can he judge of their Ability? The confequence will be, that a great part of the Army will be deftroyed in an ufelefs manner, and the General's Character will fuffer, if the Succefs does not anfwer, and not they.

On the contrary, if the Engineers are duly qualified, in all refpects, to anfwer what is expected from them, will the General know himfelf well enough to fubmit entirely to their Advice and Underftanding, and regulate his Conduct according to their Judgement? A Task more difficult to an ignorant Perfon who is brave, than to one lefs fo.

If a General, after a Defeat, is obliged to fhut himfelf up in a fortified Place, and is there attacked by a fuperior Force, it will not be in his Power to make a good Refiftance, if he is unskilled in the Manner of defending Places; all the great Expence a Prince or Nation is at in fortifying it, will not ferve to fecure the Country from Danger, if his is obliged to retire from one Place to another, fo faft as the Enemy advances; and if he undertakes to make any Refiftance, will expofe the greateft Part of his Army to Deftruction for no purpofe, and the Country ftill left to the Mercy of the Conqueror. It is not the Lofs of a Battle, an Accident the greateft General is liable to, but his Ignorance in defending himfelf in fortified Places, which he was obliged to abandon fhamefully to the Conqueror, that will be an eternal Reproach to him.

Fortifications are now become almoft as numerous as Towns, that it is impoffible either to preferve a Country or make a Conqueft, without being acquainted with the Art of Attac and Defence; wherefore it is of the greateft Importance to an Officer to be well inftructed therein, fince every

one

one from the loweſt to the higheſt has a Chance, or at leaſt Hopes, to become chief Commander one time or other; beſides, every one is liable to be intruſted, in the Courſe of a War, with ſome Poſt or other, either to ſecure Convoys, or to annoy the Enemy, or prevent them from plundering the Country by Parties, wherein if they acquit themſelves with Conduct and Prudence, it will be a ſure road to Preferment, and acquire them a laſting Applauſe.

The celebrated *Vauban* was the firſt that made any conſiderable Improvements in this Part of the Art of War; he not only deſcribed the Manner of attacking Places in general, but alſo entered into every minute Particular that may happen, during the whole Courſe of a Siege, according to the different Circumſtances, Times, and Occaſions. Before his Time ſingle Approaches only were made, fortified with a few Redouts to ſupport the Workmen. But as this way of approaching did not embrace the whole Front attacked, the Beſieged had Opportunities to Sally out and fall upon the Workmen and their Guard on every Side, drive them out of their Works and deſtroy them, by which means a Place but moderately fortified was able to withſtand an Army a conſiderable Time, and oft until they were relieved, or the Badneſs of the Seaſon obliged the Beſiegers to retire.

But ſince *Vauban* has invented the Parallels or Places of Arms, and the Ricochet Firing, to which may be added the numerous Artillery with which Armies are now provided, the Art of Attacking is rendered ſo ſuperior to that of Defending, that thoſe Places which were formerly eſteemed impregnable, can ſcarce ſtop a victorious Army a Month: And nothing

nothing but great Garrifons, and large Places well fortified, and provided with a fufficient Quantity of Artillery, and every Neceffary for a good Defence, can have any Chance for making a tolerable Refiftance. For by joining two Attacs with thefe Parallels, and the making half Parallels or Places of Arms between the fecond and third Parallels, the Befiegers can always oppofe a greater number of Troops againft the Befieged, than they can, and by that means are able to prevent Sallies from the Garrifon, or if they make any, drive them back with confiderable Lofs: And by means of the Ricochet Batteries, the Troops placed behind the Parapets are fo much expofed to the Fire of the Befiegers as if there were none.

As *Vauban* has treated this Subject in fo full and ample a Manner as could be expected, the Reader may be furprifed that I have not rather given a juft Tranflation of his Work than wrote a new one. This is what I intended at firft, but feveral Reafons have determined me to the contrary. The Order and Method perfued by that Author did not appear fo well chofen as could be wifhed; fuch as treating of the fame Subject in different Places; feveral Improvements have been made fince his Time; which, tho' not very confiderable, yet fufficiently deferve not to be neglected; he likewife fuppofed the Reader to be acquainted with the common practice of Engineering, which I prefume not to be the Cafe of our Military Gentlemen in general; wherefore particular Care is taken in this Work to omit nothing that may render the Subject intelligible even to the meaneft Capacity, fo as to make it of general Ufe to all Readers, without being obliged to have any other previous Knowledge than

that

that of underſtanding the Plans of Fortifications, without which it would be impoſſible to explain the Subject.

This Work is divided into Three Parts; the firſt contains the Attac, the ſecond the Defence, and the third the Manner of making Mines. The firſt begins with the neceſſary preparations previous to a Siege, and then every Part of the Attac is diſtinctly treated of from the Beginning to the End, with the manner of tracing the Plan of the Attacs on Paper at firſt, and thence transferring it to the Ground; the Manner of placing and making the Batteries is alſo particularly explained; in ſhort, nothing material is left untouched, which may in the leaſt conduce to the full underſtanding the Subject.

In the Second is given every Thing that relates to the Defence of Places, beginning with the Garriſon, Ammunition, and Proviſion a Place ought to be provided with in reſpect to its Bigneſs and Strength, and the neceſſary Preparations in general previous to the Siege, to put it in a Condition of reſiſting ſo long as the Nature of the Place will admit of. I have ſomewhat differed in the Eſtimate of Troops and other Neceſſaries, from M. *Vauban*, on account that the Beſiegers are now more numerous, and provided with much more Artillery than in his Time. After this are given the Particulars of the proper Defences peculiar to each Work, wherein I have inſiſted much on the Defences of Forts and Caſtles, not omitting Churches and Farm-houſes, which are ſometimes of uſe to place ſmall Parties in, either to annoy the Enemy, or to prevent their annoying the Army, or making Inroads into the Country to raiſe Contributions. It likewiſe

oft

oft happens that a Party is ſurprized by a ſuperior Force, who to prevent their being taken may be obliged to retrench themſelves in the beſt Manner they can in ſuch Places: For which Reaſon I ſuppoſe it will not be altogether uſeleſs to ſhew how ſuch Places, however inſignificant in themſelves, may reſiſt for ſome Time a ſuperior Force, eſpecially if that Force be unprovided with Artillery; and as this Kind of Party-fighting doth more frequently happen than Engagements of whole Armies, if well managed they may prove of great Advantage in the End to an Army, tho' inferior to their Enemies.

The Third and laſt Part, contains a Treatiſe of Mines, more compleat than any hitherto extant, the Principles of which have been deduced from a Multitude of Experiments, made by M. *Belidor*, formerly Profeſſor of the Royal Academy of Artillery at *La Fere*; and others, as Mr. *De Valliere*, Lieutenant and Inſpector General of the Artillery, and Mr. *Maigrigny* an old experienced Engineer. Their different Sentiments upon this Subject are candidly diſcuſſed, on the Reſult of which, and our own Experience, this Theory is formed; and to ſhew its Agreement with the Practice, ſeveral Examples are given, by which it appears to anſwer as near as can be expected from the Nature of the Subject.

M. *Vauban* has indeed treated very largely upon Mines, in his Memoir of the Attac and Defence of Places: But as in his Time the Subject was in its Infancy, and not brought to that Degree of Perfection it is now arrived at, it is not to be preſumed that he could have given all that may be ſaid of Mines. Beſide, the great Number of Experiments

periments made ſince his Time at *La Fere,* moſt of which I was an Eye-Witneſs to, has opened a much larger Field for Speculation.

Though M. *Belidor* was the chief Director of the Experiments made at *La Fere,* yet in his Memoirs upon this Subject, not yet printed, he has built his Theory upon an Hypotheſis that may with reaſon be conteſted. For notwithſtanding I had found, by meaſuring ſeveral Excavations of Mines, that their Figure is nearly a Paraboloid, he nevertheleſs rejects this, and makes uſe of quite another, whereby the Charges of Mines are Determined.

It is certain that the World is obliged to him for exploding the old Errors, that the Diameter of the Baſe is always double the Line of leaſt Reſiſtance, or the Depth of the Mine; and when a Mine is loaded with a greater Charge than neceſſary, inſtead of producing a proportional Effect, that it makes only a narrow Hole or Pit: This has been the Opinion of all the Miners and Engineers before his Time; and ſo long as this Error ſubſiſted, it was impoſſible to bring the Theory to any Perfection; ſince, according to this Notion, a Mine whoſe Diameter of the Baſe is 80 Feet, muſt be 40 Feet deep, which is not always poſſible; and when it may be done, would coſt immenſe Labour, and ſome thouſand Pounds of Powder more than is neceſſary, as I have ſhewn in this Treatiſe. This Work concludes with three Tables, the one computed by Mr. *De Valliere,* and the other two by me, and the Charges of Mines of the ſame Openings are compared, by which appears the great Advantage my Theory has above that of others.

Before I conclude, I cannot without Ingratitude omit to mention that excellent Work of M. *Le Blond,*

Blond, Profeſſor of Mathematics to the Pages belonging to the King's great Stables, called *Elemens de la Guerre,* to which I am particularly obliged for the Order in which this Subject is treated; and I have not ſcrupled to uſe ſome Paſſages of his where I found them uſeful: The Capitulation, inſerted at the End of the Defence, is entirely from his Book, which I recommend to thoſe who underſtand *French,* as being treated with more Clearneſs and Elegance than any other Work of the Kind.

CON-

CONTENTS.

How

CONTENTS.

Prin-

Principal Terms used in this Book Explained.

TO *Besiege a Place,* is to surround it with an Army, and approach it by passages made in the ground, so as to be covered against the fire of the place.

When an Army can approach so near as the covert-way, without breaking ground, under favour of some hollow roads, rising grounds, or cavities, and there begin the work, it is called, *Accelerating the Siege.*

And when the Army can approach the town so near as to take it, without making any considerable works, the Siege is called an *Attack.*

To Block up a Place, is to surround it with a sufficient number of troops, so as to prevent any succour or provision being thrown into it.

The intent of a Blockade is, to *reduce* a garrison *by famine.*

A place should never be blockaded, but on a certainty that there is but a small quantity of provision in it, otherwise it would require too much time, and, perhaps, not be taken at last.

To invest a Place, is to surround it with troops, as in a blockade, so as to prevent any succours, *&c.* It is properly a *preparative for a Siege.*

To Insult a Work, is a sudden and unsuspected attack, with small arms, or sword in hand.

Surprize, is to take a Place by stratagem or treason.

To Escalade a place, is to approach it secretly, and to place ladders against the wall, or rampart, for the troops to mount and get into it that way.

To

To Petard a place, is privately to approach the gate, and fix a petard to it, ſo as to break it open for the troops to enter.

Line of Circumvallation, is a kind of Fortification, conſiſting of a *parapet,* or breaſtwork, and a ditch before it, to cover the Beſiegers againſt any attempt of the enemy in the field.

Line of Countervallation, is a breaſtwork with a ditch before it, as the line of circumvallation, to cover the Beſiegers againſt any ſally from the garriſon.

The Works made to cover an Army, ſo as to command a part of the country, with a breaſtwork and a ditch before it, are alſo called *Lines.*

A Work made round the camp of an Army, to cover it againſt any ſurprize, is called a *Retrenchment.*

An Epaulement, is a breaſtwork of eight or ten feet high, to cover the cavalry. The breaſtwork of a battery is alſo called an *Epaulement.*

Trenches, is a general term for all the works ſunk in the ground, and made during a Siege, or in any Attack.

Approaches, are thoſe trenches which are carried on towards the place in *Zig-zags,* or turnings, ſo as not to be *enfiladed* by any works of the place.

Parallels, or *places of Arms,* are thoſe trenches which join both Attacks, and are, as it were, parallel to the front works of the place.

There are generally three in an Attack of a place.

Saps, are trenches made and carried on under cover with gabions and faſcines on the flank, and mantlets, or ſtuffed gabions, in the front.

They

They differ therefore from the other Approaches, in being carried on under cover, and in not being ſo wide.

The Rear of an Attack, is where it begins; and the *Front* or *Head,* that part next to the place.

Breach, is the opening made in the rampart of a work, either by cannon or mines.

Aſſault, is a vigorous and ſudden Attack, made in the breach, to get maſter of the work.

Lodgement, is a kind of trench or retrenchment made in a work, after the enemy is drove out, to cover the troops from the fire of the place.

PLATE I. There are many different kinds of materials uſed on theſe occaſions, as *Faſcines, Pickets, Gabions, Sand-bags, and Woolpacks.*

Faſcines, are compoſed of branches like fagots, of ſix feet long, and eight inches in diameter, tied in two places.

There is another ſize of faſcines, three feet long only, called *Sap fagots.*

And when a Faſcine is made from twelve to ſixteen feet long, it is then called *Sauciſſon,* and is uſed in making of batteries.

Pickets, are ſtakes of three feet long, and about an inch and a half diameter, pointed at one end.

Gabion, is a kind of basket without a bottom, every where equally wide, three feet high, and as much in diameter, having nine or ten ſtakes of about an inch and a half diameter; which exceed the basket-work five or ſix inches, and are pointed, and generally made by the miners and ſappers, who underſtand this kind of work.

Sand-Bags, are ſacks of two feet high, and ten inches diameter, filled with earth, and tied very faſt.

Wool-

Wool-packs, are five feet high, and about fifteen inches diameter, filled with wool; they ſerve to cover the troops, as ſoon as they have got poſſeſſion of a work, till the lodgment is made.

Beſides theſe materials, there are *Blinds, Chandeliers, Mantlets, Stuffed gabions, Chevaux de frize*, and *Crows-feet.*

A Blind, is a wooden frame of four pieces, round or flat, two of which are ſix feet long, and pointed at the ends, and the others three or four feet, which ſerve as ſpars to faſten the two firſt together.

Their uſe is, to fix them upright in the ground, againſt the ſides of the ſaps, to ſuſtain the earth, and to faſten faſcines on the upper part, or to cover the ſap, and to lay faſcines over them, to ſecure the troops from ſtones and granades.

Chandeliers, are wooden frames, made of two pieces fixed croſs-ways on two other pieces, at about four feet aſunder, and upon their interſections are erected two vertical pieces of five feet high, each ſupported by three buttreſſes; the interval of theſe two pieces is filled up with faſcines, to cover the troops upon occaſion.

Mantlet, is a wooden fence, rolling upon wheels, of two feet diameter; the body of the axel-tree is about four or five inches ſquare, and four or five feet long, to which is fixed a pole of eight or ten feet long, by two ſpars; upon the axel-tree is fixed a wooden parapet three feet high, made of three inch planks, and four feet long joined with dowel-pins and two croſs-bars; this parapet leans ſomewhat towards the Pole, and is ſupported by a brace

brace, one end of which is fixed to the pole, and the other to the upper part of the parapet.

The ufe of the mantlet is, to cover the fappers in front againft musket-fhot.

A ftuffed Gabion, is a gabion of five or fix feet long, and as wide, ftuffed quite full with all kind of fmall wood, or branches; they are ufed inftead of mantlets, and for the fame purpofe.

Chevaux de Frize, is a piece of timber, either round or cut into feveral faces, through which are drove ftaves of about an inch and a quarter in diameter, and fix feet long, pointed at the ends, fo that lying any way, it always prefents fome points.

Their ufe is, to prevent the horfe from breaking in upon the foot.

Crows-feet, is a kind of iron with four points, fo difpofed as always to have three points downwards and one upwards; they ferve alfo to ftop the horfe.

The figures of all thefe materials and machines are reprefented in the firft plate; which, together with what has been faid, may be fufficient to have them underftood.

The principal terms being explained, the others, which are ufed hereafter, will alfo be defined in their proper places.

Lately

Plate I.

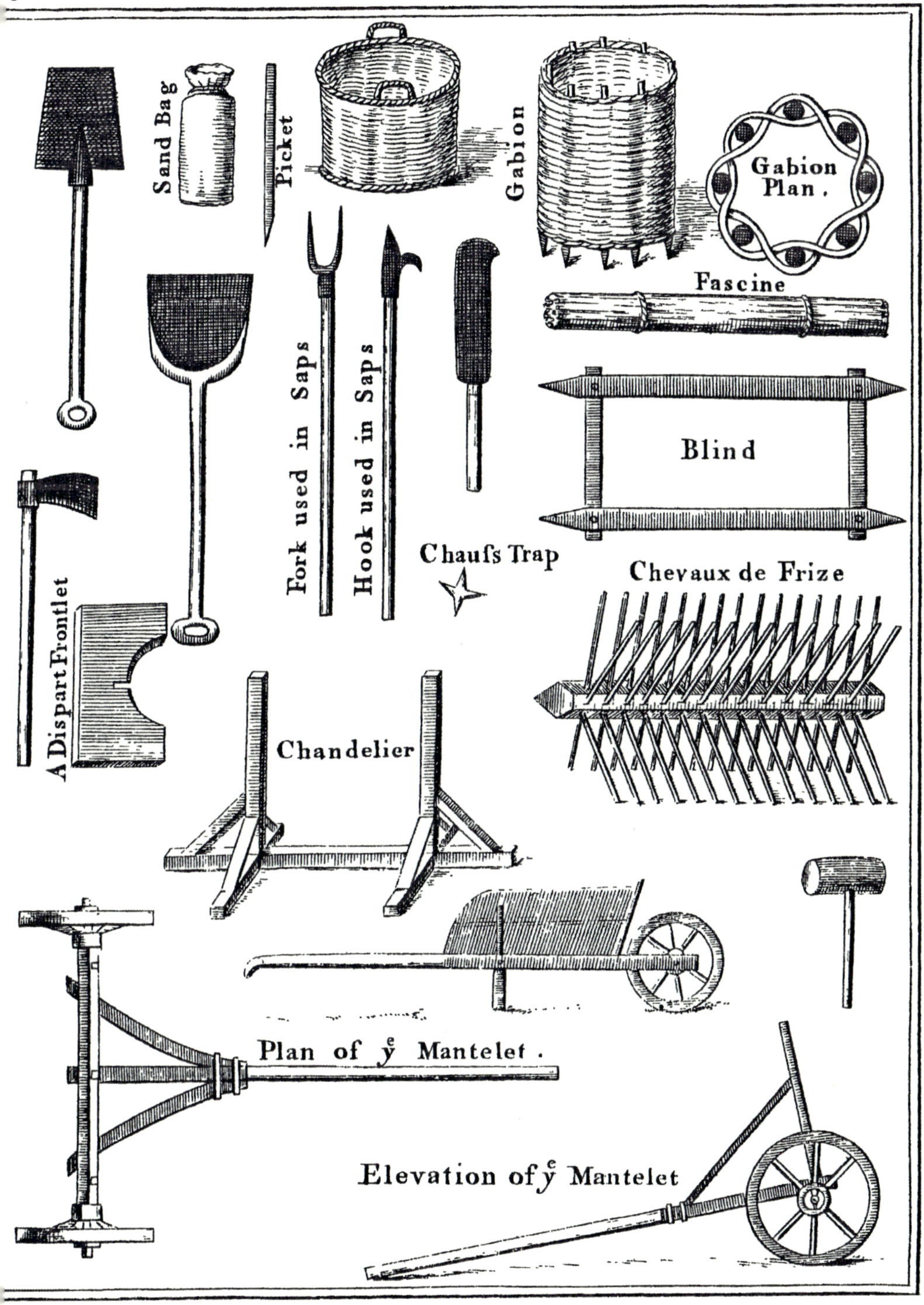
Sand Bag
Picket
Gabion
Gabion Plan.
Fascine
Fork used in Saps
Hook used in Saps
Blind
Chaufs Trap
Chevaux de Frize
A Dispart Frontlet
Chandelier
Plan of ye Mantelet.
Elevation of ye Mantelet

Lately publiſhed, by Mr. MULLER, Price 7 s. 6 d.

I. A TREATISE of ARTILLERY: Containing, 1. General Conſtructions for Braſs and Iron Guns, for Land and Sea; with their Carriages, Mortars, and Howitzes; their Beds and Carriages, and Dimenſions of all other Kind of Carriages; Exerciſe of the Royal Regiment of ARTILLERY, and its Service at Home and Abroad; in a SIEGE or BATTLE; MARCH, INCAMPMENT, AMMUNITION, STORES, and HORSES, LABORATORY-WORK, THEORY of POWDER applied to FIRE-ARMS.

II. A SYSTEM of MATHEMATICKS, FORTIFICATION, and ARTILLERY. In Six Volumes, Illuſtrated with One Hundred and Forty large fine Copper-plates. Price 1 l. 16 s.

Vol. I. Algebra, Geometry, and Conic-Sections. II. Trigonometry, Surveying, Levelling, Menſuration, Laws of Motion, Mechanics, Projectiles, Gunnery, *&c.* Hydroſtatics, Hydrawlics, Pneumatics, and Theory of Pumps. Twenty-four Cuts. III. Fortification, Regular and Irregular 33. IV. Practical Fortification in four Parts 26. V. Artillery in ſix Parts 29. VI. Attack and Defence of Fortified Places, Mines, *&c.* Three Parts, 28 Cuts.

III. A SYSTEM of CAMP-DISCIPLINE; Military and Naval Honours, Garriſon Duty, Regulations for Land and Sea Officers, *Kane*'s Campaigns of K. *William* and the D. of *Marlborough*, continued to 1757. By an Impartial Hand.

IV. LIST of the ARMY, with a Succeſſion, 3 s. 6 d.

V. *Millan*'s Univerſal Regiſter of Court and City Offices, Army and Navy, *&c.* 2 s. 6 d.

VI. *Millan*'s Signals, Flags, Coins, Weights, *&c.* 9 s.

ATTAC

ATTAC and DEFENCE OF FORTIFIED PLACES.

PART I.

IN Sieges, as in all undertakings, it is neceſſary timely to conſider every minute circumſtance that may happen during the execution of the deſign, and to provide in the beſt manner againſt every ſeeming obſtacle. Not only the General's character, the reputation of the Army, and the glory of his country, are concerned, but a prodigious expence muſt unavoidably attend ſuch an exterprize; all which are entirely loſt in caſe of a miſcarriage, beſides the lives of a number of men, a more ſenſible loſs to the nation.

For theſe reaſons, a General, before he reſolves on a Siege, ſhould exactly know, 1ſt, the ſtrength of the enemy, as well in the field as in the adjacent garriſons; and, 2, what number the enemy may draw from other quarters during the Siege; 3, the ſtrength of the place, its ſituation, and nature of the ground about it; 4, the ſtrength of the garriſon, and goodneſs of the troops therein, together with the quantity of artillery, ammunition, ſtores and proviſion in the place; 5, if the works are countermined or not, and whether any part of the

the country about the placè may be laid under water, by means of dykes or ſluices; laſtly, he ſhould be acquainted with the skill and bravery of the governor, engineers, and the reſt of the officers.

Theſe particulars known, he ſhould conſider whether he is maſter of the field, and able to repulſe the enemy, if he was attacked during the ſiege, without leaving the trenches unprovided, and to reſiſt, at the ſame time, any ſally the garriſon ſhould make; and whether, in caſe the enemy ſhould attack one of his towns during the ſiege, he may have time to relieve it after the ſiege is over, or how to provide otherwiſe againſt ſuch an accident.

He ſhould alſo know whether he has a ſufficient number of engineers, and their capacities; whether he has or can provide a ſufficient quantity of artillery, ammunition, ſtores and proviſion, with the number of horſes for draught or carriage; whether the commandant of the artillery be well qualified for his place, what funds will be wanting, and how they may be ſupplied; in fine, he ought to be appriſed of every thing neceſſary for ſuch an important undertaking.

As the ſucceſs will redound chiefly to the honour of the General, as well as the miſcarriage be laid to his charge, he ought therefore to be thoroughly acquainted with theſe particulars, without being obliged to truſt to the judgment or capacity of others.

STORES.

THE quantity of ſtores required for a Siege cannot be preciſely determined, on account of the various conſiderations on which it depends; ſuch

ſuch as the ſtrength of the place and garriſon, the capacity of the governor and engineers, the quantity of artillery, ammunition, ſtores and proviſion; laſtly, on the ſeaſon, place, ſituation, *&c.* But as it is neceſſary to give ſome idea to the unexperienced officer, I ſhall here ſet down the quantity of each kind, for a month's Siege, as eſtimated by Marſhal *Vauban,* whom we chuſe to follow, on account of his great experience and undoubted judgment.

Stores required for a Month's Siege.

POWDER, as the garriſon is more or leſs ſtrong,		8 or 900,000 lb.
Shot	for battering pieces —	6,000
	of a leſſer ſort, — —	20,000
Battering cannons, — — —		80
Cannons of a leſſer ſort, — —		40
Small field-pieces for defending the lines,		20
Mortars for throwing	ſhells, ——	24
	ſtones, ——	24
Shells for mortars, ——		15 or 16,000
Hand granades, — —		40,000
Leaden bullets, —— —		180,000 lb.
Matches, — ——		10,000 braces
Flints for muskets, beſt ſort, —		100,000
Platforms compleat for guns, ——		100
Platforms for mortars, —— ——		60
Spare	carriages for guns, ——	60
	mortar-beds, —— ——	30
	ſpunges, rammers and ladles,	20 ſets
Tools to work in the trenches, ——		40,000

Several hand-jacks, gins, ſling-carts, travelling-forges, and other engines proper to raiſe and carry heavy burdens, as likewiſe ſome to carry water to extinguiſh fire.

Several

Several parcels of ſpare timber, for bridges, wheelwrights, carpenters, &c.

There are beſides ſeveral other things neceſſary, as miners tools, mantlets, ſtuffed gabions, faſcines, pickets and gabions, in great quantities; tools for ſmiths, carpenters and wheelwrights; a number of horſes for the artillery; carts and waggons, ſuch as can be procured in the country, are allſo uſed upon occaſion.

INVESTING PLACES.

PRe-requiſite to inveſt a place with ſucceſs, the General ſhould uſe various ſtratagems to deceive the enemy, and prevent their gueſſing his real deſign; ſometimes the deceit may be carried ſo far as to inveſt another place; at other times marching with the army, as if intended to attack the enemy, in order to drive them at ſome diſtance from the place, and then return quickly to inveſt it. In ſhort, no opportunity ſhould be neglected to arrive before the place, before the enemy has either time to throw in troops, ammunition, or proviſion, ſince the ſucceſs of the Siege depends chiefly on this diligence.

The place is inveſted in the following manner.

A Body of 4 or 5000 horſe is detached from the army, if the country is open, or a body of horſe and foot, full of defiles and woods, commanded by a Lieutenant-General, and two or three Major-Generals, who march with all poſſible ſpeed, day and night, till they come within four or five miles of the place, where they halt, to conſult, and divide themſelves into as many parties as there are principal avenues leading to it; then they march on, ſo

as

as to arrive in the dusk of the evening at their ſeveral appointed poſts, much about the ſame time; which ſhould be juſt out of the reach of cannon-ſhot from the place.

This done, ſmall parties are ſent to the very gates to carry off men, cattle, and whatever may be ſerviceable to the garriſon. Theſe parties are to be ſupported by ſome ſquadrons of horſe; and it would not be amiſs, if they received ſome cannon-ſhot, to diſcover the reach of their guns. In the mean time, the reſt of the detachments take their poſts in the moſt convenient places, ſo as to prevent ſuccours being thrown into the town.

In the day-time they keep without cannon-ſhot; but as ſoon as it is dark, the ſeveral parties approach the place as near as poſſible, ſo as to leave but ſmall intervals between them; then turning their backs upon the town, and placing guards before and behind them, to prevent any ſurprize, half the troops are to keep always mounted, whilſt the reſt refreſh.

At Dawn they retire by degrees, obſerving the ſituation of the place, the nature of the works, and the ground round it, till they come to the former poſts, where they place proper guards towards the town, and in all the principal avenues towards the country; the reſt repoſe, keeping their horſes ſaddled for mounting on a minute's warning.

Parties are ſent to reconnoitre the enemy, while the commanding officer and engineers pitch upon a proper place for encamping the army, as ſoon as it arrives, and obſerve where the line of circumvallation is to be made.

The day the place is inveſted, the artillery begins to march, with all the ſtores and ammunition neceſſary

neceſſary for a Siege, whilſt, on the other hand, the army makes forced marches, and arrives commonly within three or four days after the inveſting.

The commander of the detachment goes about two or three miles to meet the General, with an account of his proceedings, on which the General ſettles the diſpoſition of the camp.

The next day he rectifies what miſtakes may have happened, and goes to reconnoitre the place, attended by the reſt of the general officers, and chief engineers, ſo that the ſituation of the line of circumvallation may be determined.

This done, the encampment regulated, and the troops placed in the order agreed on, the General aſſigns to the other general officers their quarters; the chief or head quarter is fixed upon, as alſo thoſe for proviſion, and the park of artillery. All theſe particulars are to be rectified, ſo ſoon as the place for opening the trenches is determined.

In the mean time, ſmall guards are poſted near the town, in the moſt convenient places, ſuſtained by larger, to ſtreighten the garriſon as much as poſſible, and the engineers roughly trace the line of circumvallation, with rods and pickets only, in order to regulate the encampment.

Diſpoſition and Manner of making the Line of Circumvallation.

PLATE II. IN the tracing this line, regard muſt be had to the following particulars. 1ſt, to occupy the moſt convenient ſpot of ground about the place, though it be ſomewhat nearer or farther from the town, than it ſhould otherwiſe be. 2d, to preſerve ſuch a diſtance, that the camp may be be without the reach of cannon-ſhot from the covert-way, which is reckoned to be about 1200 fathoms. 3d, Not to keep a greater diſtance than is neceſſary for the ſecurity of the camp. The depth of the camp is generally 30 fathoms, and the diſtance between the front of the camp and the line 120. 4th, To avoid all high places which might command any part of the camp, and when that cannot be done, to take them within the line; but if the line ſhould thereby become too extenſive, they muſt be fortified with redoubts, or other outworks.

All advantages ariſing from the nature of the ground are likewiſe to be made uſe of; ſuch as precipices, moraſſes, rivers, brooks, hedges, and, where it can be done, felling of trees.

The dimenſions of the ſeveral parts of the line of circumvallation are, 1. The diſtance from the point of one *Redan* to that of the next, is commonly 120 fathoms; ſometimes 10 or 12 fathoms, more or leſs, according as the nature of the ground will permit. 2. The openings or gorges of the *Redans* about 30, and the capital or depth 20: The *Redans* are to be placed on the higher part of the ground. As to the reſt, the form and ſituation of the line is always adapted to the nature of the ground, and, provided it is every where well flanked, it will be ſufficient.

The gorgers of the *Ravelins,* which cover the gates or ſally-ports, are thirty fathoms, the capitals twenty, and the flanks ſix: The ditch before the line is from fifteen to eighteen feet wide above, but reduced to one-third below. Hence a ditch of eighteen feet wide above, will be but ſix feet at the bottom; and if its depth is ſeven feet and a half,

half, the excavation of fix feet, or one fathom in length, will be fifteen feet, or two cubic fathoms and a half, which is reckoned to be as much work as one man may conveniently do in feven days.

According to this computation, I fhall infert the meafures of fix different profils given by M. *Vauban*, which may ferve on all occafions.

	PARAPET.				DITCH.			
Prof.	Thickn.	Heighth within.	Heighth without.	Width above.	Width below.	Depth	Solid Cont.	Days Work
I	8 : 0	7 : 6	6 : 0	18:0	6 : 0	7 : 6	15 : 0	7 Days
II	7 : 0	7 : 6	6 : 0	16:0	5 : 4	7 : 0	12 : 5	6 : 18
III	6 : 0	7 : 6	6 : 0	14:0	4 : 8	6 : 6	10 : 1	5 : 16
IV	6 : 0	7 : 6	6 : 0	12:0	4 : 0	6 : 0	8 : 0	4 : 0
V	5 : 0	7 : 6	6 : 0	10:0	3 : 4	5 : 6	6 : 1	3 : 0
VI	4 : 0	7 : 6	6 : 0	8 : 0	2 : 8	5 : 0	4 : 5	2 : 0

N. B. All dimenfions are expreffed in feet and inches, as well as the folid contents, fix feet of which make a cubic fathom, 12 inches a foot, fuppofing the length of the excavation to be fix feet, as mentioned above *.

Suppofing a cannon-fhot from the covert-way to reach 1200 fathoms, allowing 200 more to the rear of the camp, makes 1400; then taking 30 for the depth of the camp, and 120 more for the diftance from the front of the camp to the line, the whole includes

* The method of computing the folid content is, to multiply half the fum of the width above and below of the ditch, by its depth, and divide the product by fix. See my *Elem. Mathem.* p. 247.

includes a diſtance of 1550 fathoms between the line and covert-way.

Now if the Place to be attacked is a regular *Octagon,* fortified according to M. *Vauban*'s firſt Method, the diſtance from the center of the place to the covert-way, will be about 250 fathoms, which being added to 1550, gives 1800 fathoms for the diſtance of the line from the center of the place; and the circumference of the Radius 1800 will be 11314.

Therefore the line of circumvallation would be 11314 fathoms, if it was circular; but on account of its turnings and redans, there may be added 1880, making together 13194 fathoms, which being divided by 2282 (the number of fathoms or toiſes contained in a *French* league) gives about $5\frac{2}{3}$ leagues or 17 *Engliſh* miles, for the whole extent of the line.

The parts of the line moſt expoſed to be inſulted, are paliſſaded, and ſometimes out-works are added, or a ſecond ditch; and when any particular part is more in danger than the reſt, holes or pits are made chequerwiſe of 5 or 6 feet deep, and about 8 feet wide above, with ſtakes planted in the middle of them, projecting about a foot above the ſurface of the ground, to prevent the cavalry from breaking in that way.

In general, the Line ſhould be every where equally ſtrong; and therefore, if ſome parts are ſtrong by nature, the reſt ought to be made ſo by art.

As the tracing the Line on the ground admits no manner of difficulty, there is no occaſion to ſay any thing of it.

As faſt as the Line is traced out, the ground is diſtributed amongſt the troops and peaſants, if any are

are to be had; this is done equally both to horſe and foot, no body being excuſed from this kind of duty.

The Sub-engineers divide the work between them; they are to take care that the work is well performed, and the dimenſions rightly followed; and the General Officers obſerve that the Engineers before their own quarters perform their duty: In the mean time, the chief and ſub-directors are employed in reconnoitering the works of the Fortification, and the ground about it; of which they are to give an account to the General, that he may fix upon the place of Attacs. The finiſhing the Line generally takes up 8 or 10 days.

BRIDGES *of* COMMUNICATION.

AS the camp is generally divided by rivers, brooks, or marſhes, it is of great importance to make a ſufficient number of bridges or cauſeways, that the communications between the ſeveral parts of the camp may be quite open and free, the neglect of which * has many times occaſioned the raiſing of the Siege, with the deſtruction of a great part of the Army.

Theſe bridges ought to be made as firm and ſecure as poſſible; there ſhould at leaſt be four of them between each of the ſeveral quarters, at about 40 or 50 fathoms diſtance from one another; their

* The *French* found by woeful experience the truth of this at the Sieges of *Turin* and *Valencienes*; at the former, tho' they were 72000 ſtrong, Prince *Eugene* attaced them with 24000 men only in one of the Quarters ſeparated by rivers from the others, and where the bridges were neglected, defeated the Army, and raiſed the Siege; and at the latter, half their Army, commanded by Marſhal *La Ferté*, and ſeparated from the other by a river, was entirely routed, and the Siege likewiſe raiſed.

their avenues of eaſy acceſs, and guards placed near them, to ſee that they are conſtantly kept in repair. Theſe bridges are always placed within the Line of circumvallation, ſo that thoſe who paſs and repaſs over them, may be ſecure, and out of all danger from the enemy.

GATES *and* SALLY-PORTS.

THE Gates are made in the principal avenues and great roads, covered with ravelins, and ſometimes with ſmall horn-works, when the Line may be eaſily attacked on that ſide: and in every ſecond curtain is made a Sally-Port of 24 feet wide, which is ſhut up by an horizontal beam turning upon a ſpindle poſt, which ſtands in the middle of the gateway.

PARK *of* ARTILLERY.

WHilſt the Line of circumvallation is making, the commander of the Artillery is employed in making the park, which ought to be placed in the part moſt remote from the town, and leaſt expoſed to be inſulted, but at the ſame ſide where the attac is to be made; it is made in the form of a ſquare redout, with a breaſtwork and a ditch round it, and the entrance is covered with a redan, or ſmall ravelin.

Beſides this, which for diſtinction is called the Great Park, and which contains all the Artillery, ammunition and ſtores of the whole Army; there is alſo a ſmaller one, at the rear of each attac, to hold the Artillery and ſtores which are uſed daily in the trenches, and which are continually ſupplied from the great Park. Theſe latter need

need not be ſo well ſecured as the great one; provided they are free from cannon-ſhot, it is ſufficient.

I ſhall hereafter explain at large, how this Park is to be formed, the ſtores are to be diſpoſed, and the quantity of each ſort required.

LINE *of* COUNTERVALLATION.

THIS Line is of the ſame nature, and made in the ſame manner, as the Line of circumvallation; as the former is intended to ſecure the Army againſt any attempt that the enemy may make from the field, ſo this latter is made to ſecure it againſt thoſe from the garriſon. For which reaſon, it is made between the camp and the place, and the redans, together with the ditch, are on that ſide next to the place. This Line is not made ſo ſtrong as the former, neither are the redans ſo near each other; and therefore the 6th profil may be uſed for it.

There are few Lines of countervallation made now, becauſe the Beſiegers are generally ſo ſuperior to the Beſieged, that they are under little apprehenſion from them. The Line of circumvallation has even of late been neglected, when the Army of the Beſiegers has been * much ſuperior to any that could come to relieve the place.

PREPARATIONS *for the* ATTAC.

AS ſoon as the Army is arrived, and encamped, faſcines, pickets, and gabions are got ready, againſt the time of opening the Trenches; they are brought

* The *French* made no Line of circumvallation in the late Siege of *Tournay*.

brought to the camp by 2 or 3000, a battalion, and 12 or 1500 a ſquadron, and placed in the fronts of their reſpective camps.

There are alſo ſeveral other things to be provided, as ſpades, ſhovels, pick-axes in great numbers, iron hooks, and forks with long poles fixed to them, for ranging and ſetting the gabions; mallets, wheel-barrows, hand-baskets and ſand-bags in great quantities; as likewiſe mantlets, ſtuffed gabions and ſauciſſons.

All theſe things muſt be ready and at hand, before the Trenches are opened; alſo the Guns mounted upon their carriages, the mortars upon their beds; the materials for making the batteries and the platforms muſt alſo be in readineſs, together with every thing elſe required in the Attac.

In the mean time, the Director of the Siege continues reconnoitering the different works of the Fortification, as likewiſe the ground about it, in order to make the plan of the Attacs.

There are a great many things to be obſerved in the tracing of the Attacs; it requires undoubtedly, the utmoſt skill and induſtry of an Engineer to fix on a proper place, and to adapt the works to the nature of the ground, the neglect of which has often been the occaſion that places were attacked on the ſtrongeſt ſides; and ſometimes the Beſiegers were obliged to change the place of their Attacs after much time and labour ſpent.

To give the reader a clear idea of the tracing an Attac, we ſhall ſuppoſe here a regular octagon, with all the ſides equally fortified, ſo that it is indifferent which of them is attacked, and endeavour to explain the general rules or maxims whereby they are conſtructed, in as plain and full a manner

as

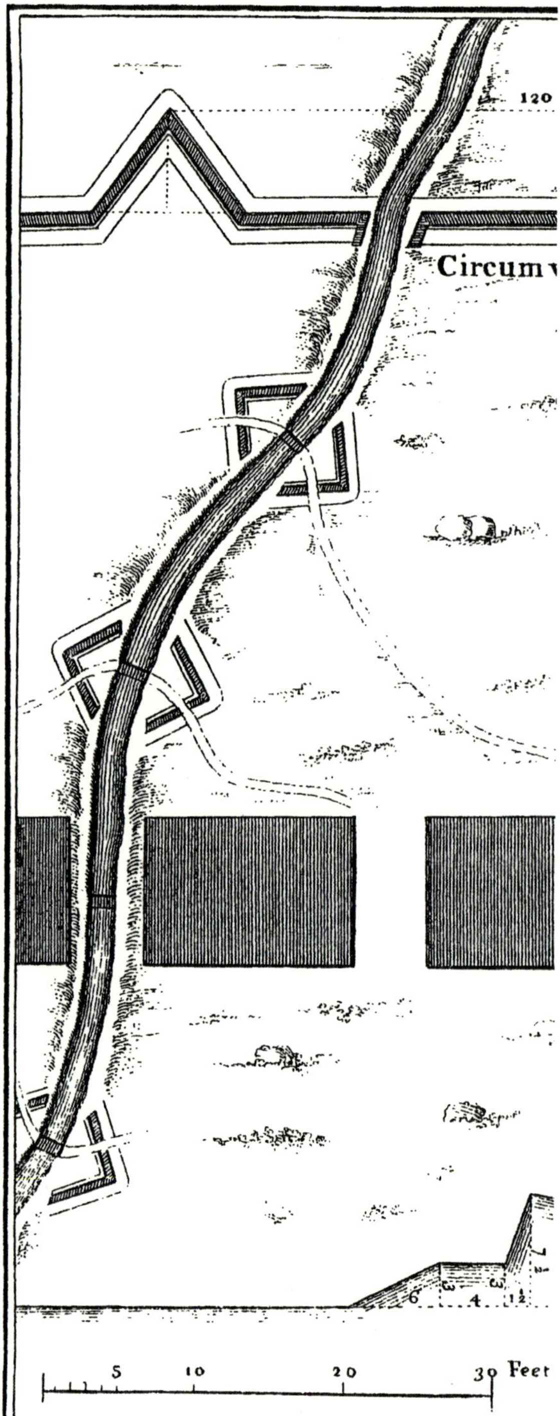
120
Circum
6
4
1½
5
10
20
30 Feet

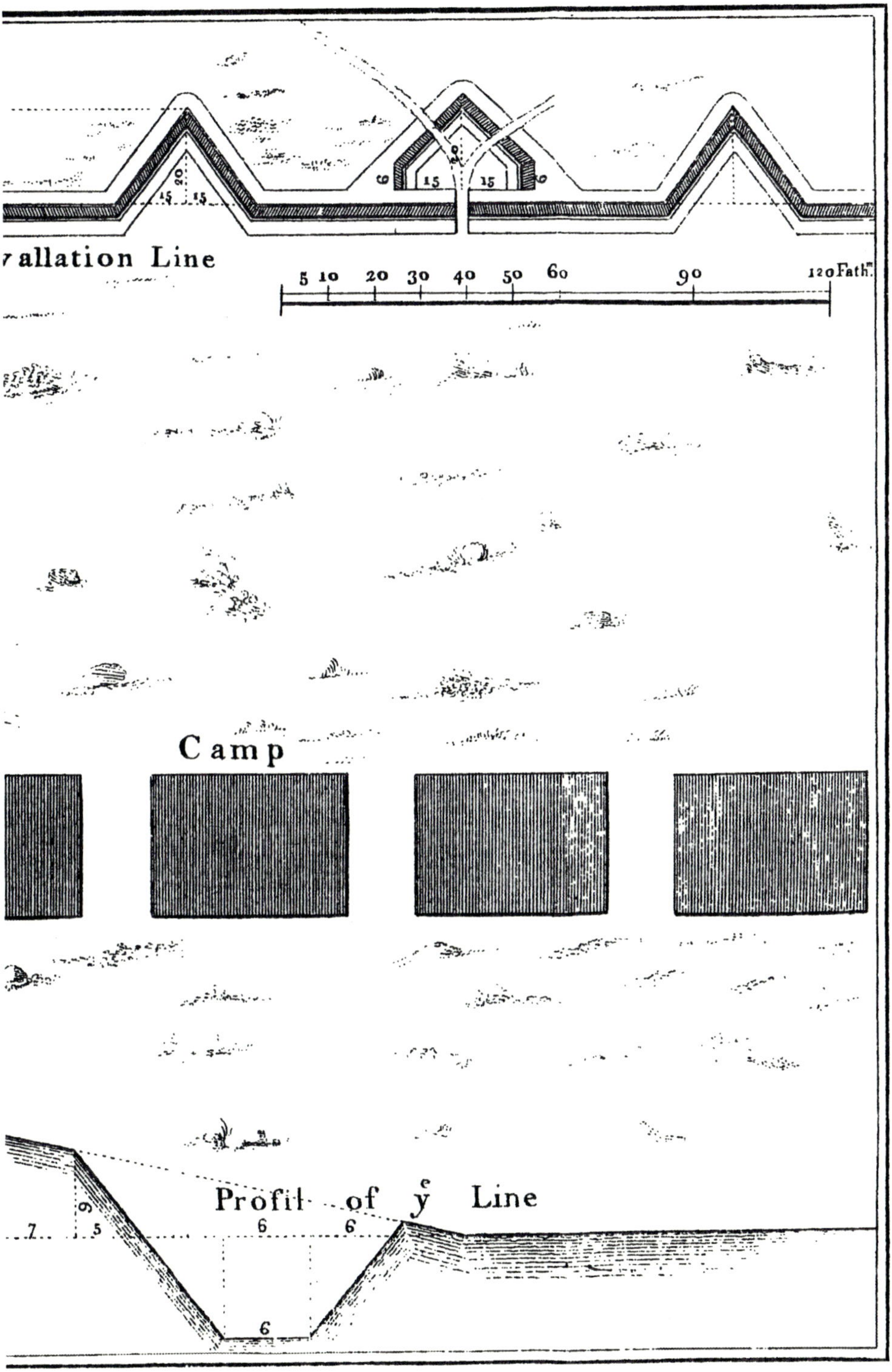

vallation Line
5 10 20 30 40 50 60 90 120 Fathm.
Camp
Profil of ye Line

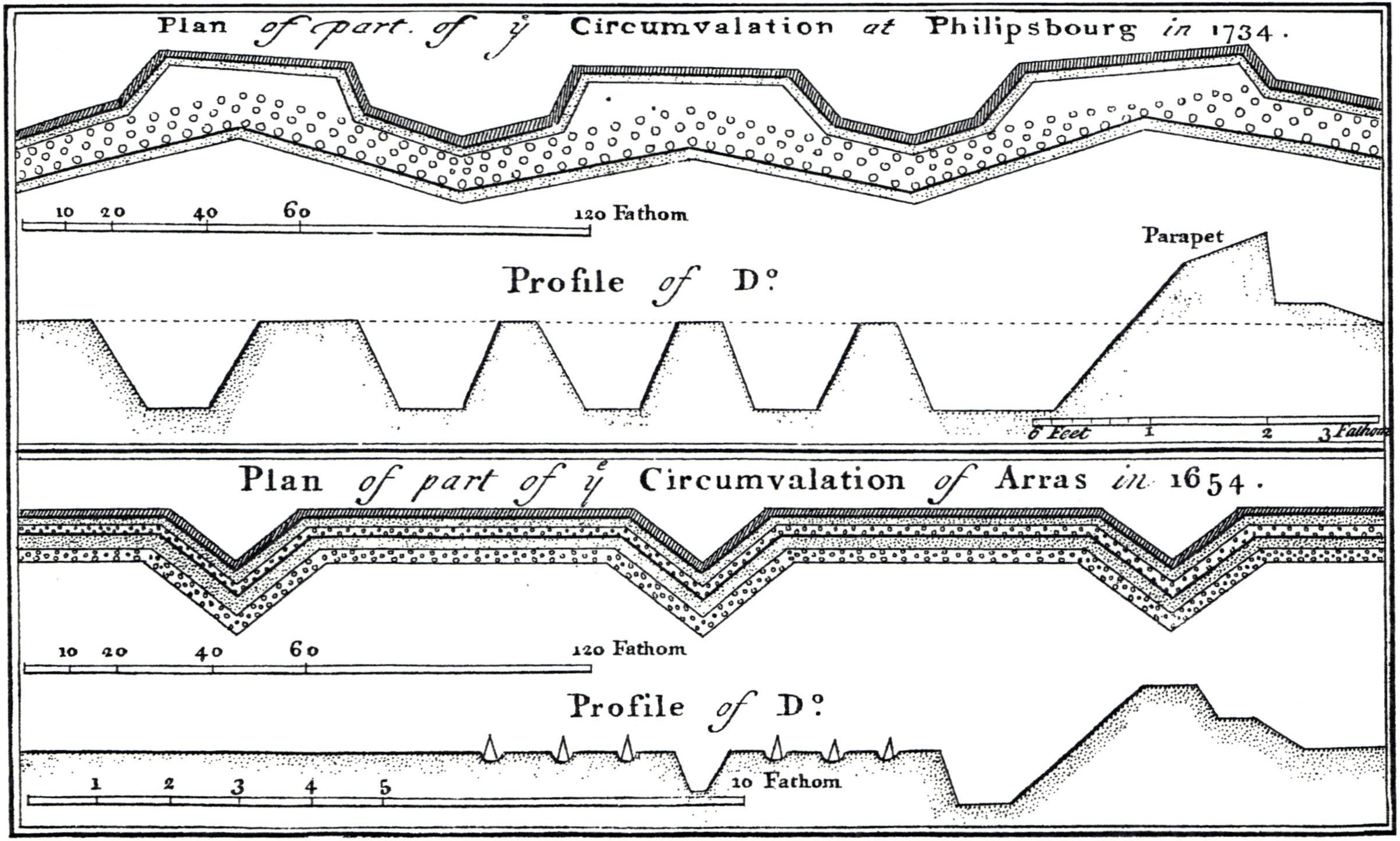

Hulett sculp.

as poſſible; which being well underſtood, will leave, I hope, no manner of doubt in the reader's mind, eſpecially when applied hereafter to irregular places.

To CONSTRUCT *the* ATTACS *on* PAPER.

PL. IV. LET, A, B, be the baſtions to be attacked, whoſe capitals, as well as that of the ravelin C, being indefinitely produced towards the field, ſet off 300 fathoms from the ſaliant-angles r, s, of the covert-way to the points L, M; and from the center O of the place deſcribe an arc of a circle through the points L, M, whoſe interſections with the capitals will terminate the ſeveral parts of the firſt Parallel P L M R; which is terminated at 20 fathoms beyond the faces produced of the two adjacent ravelins x, z, next to the front attacked. Take L N, M Q, each equal to 140 fathoms, and deſcribe from the center O another part of a circle through the points N, Q; and its interſections with theſe capitals, will terminate the ſeveral parts of the ſecond Parallel n N Q m, which extends ten fathoms only, beyond the faces of the ravelins x, z. Laſtly, draw lines through the ſaliant-angles of the glacis, and they will give the third Parallel S T. This being done, take the points D, E, in the capitals of the baſtions, at about 800 fathoms from the ſaliant-angles r, s, of the covert-way, and join them by a right line, which will expreſs the beginnings of the Attacs. From the point F in D E, at about 24 or 26 fathoms from the point E, draw a line F G, of any length, as 120 or 126 fathoms, for the firſt Approach, but ſo as when produced to fall 10 or 12 fathoms

fathoms from the fartheſt ſaliant-angle of the covert-way; from the point G draw G H, ſo as to fall as much from the fartheſt ſaliant-angle on the other ſide of the place; and the point H is taken ſo as to be nearly at the ſame diſtance from the capital s E, as the point G; and from the point H draw the line H I, ſo as to fall again as far from the fartheſt ſaliant-angle on the firſt ſide, and that the point I be equally as diſtant from the capital s E, as the point H, continue thus the Approaches to the third Parallel.

The Approaches are traced after the ſame manner on the capital D r, as alſo on the capital of the ravelin C, from the ſecond Parallel to the third; always obſerving that no part of them is to be ſeen or enfiladed from any part of the Fortification, nor to defile too much, which would occaſion more work than is neceſſary.

As the Approaches advance nearer the place, they become ſhorter, as making then ſmaller angles with each other. Between the ſecond and third Parallels, at the returns of the Approaches, are made places of arms, ſuch as X, about 30 or 40 fathoms long, nearly parallel to the third Parallel, which ſerve to receive the troops that are to cover the workmen till the third Parallel is quite finiſhed. It is to be obſerved, that as the Approaches draw near the place, the more frequent the ſallies of the Beſieged will be; and therefore it is neceſſary to have troops as near at hand as poſſible, otherwiſe the workmen would take to their heels upon the firſt alarm, and the work be much retarded.

The principal or outlines of the Trenches being thus traced, parallels are drawn to them, on the

ſide

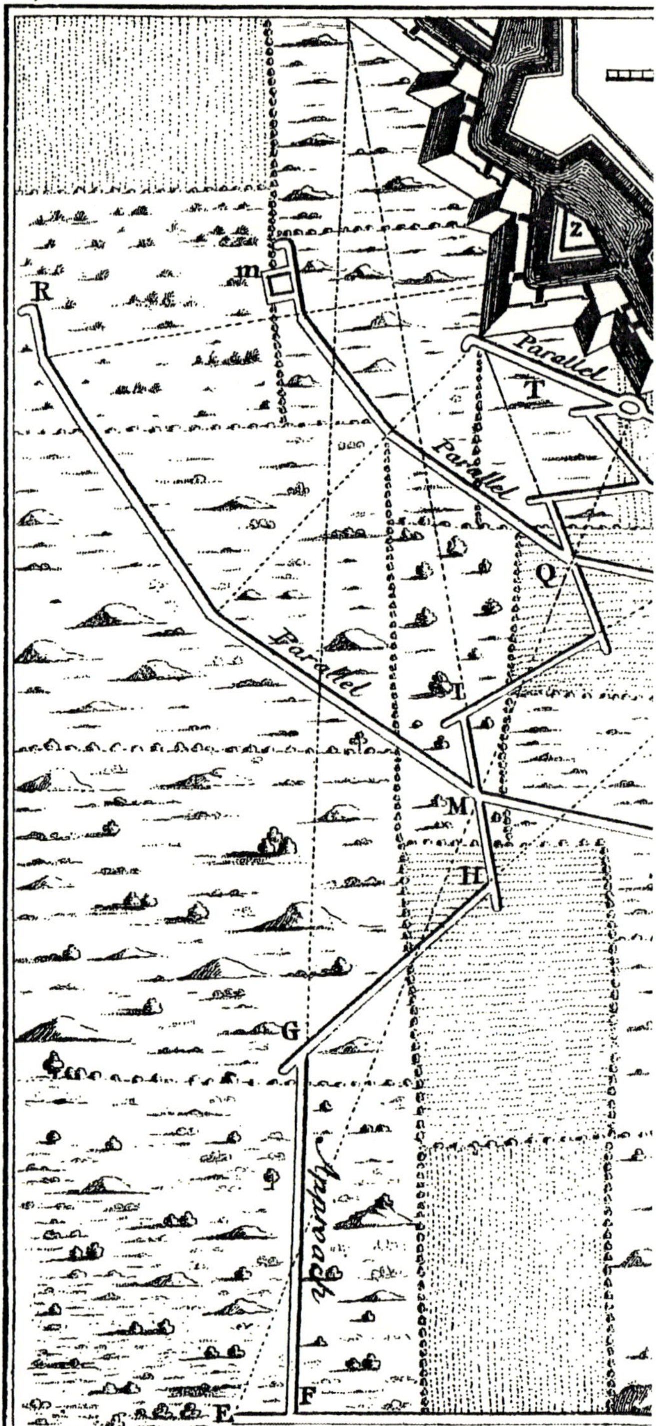

Müller Delin.

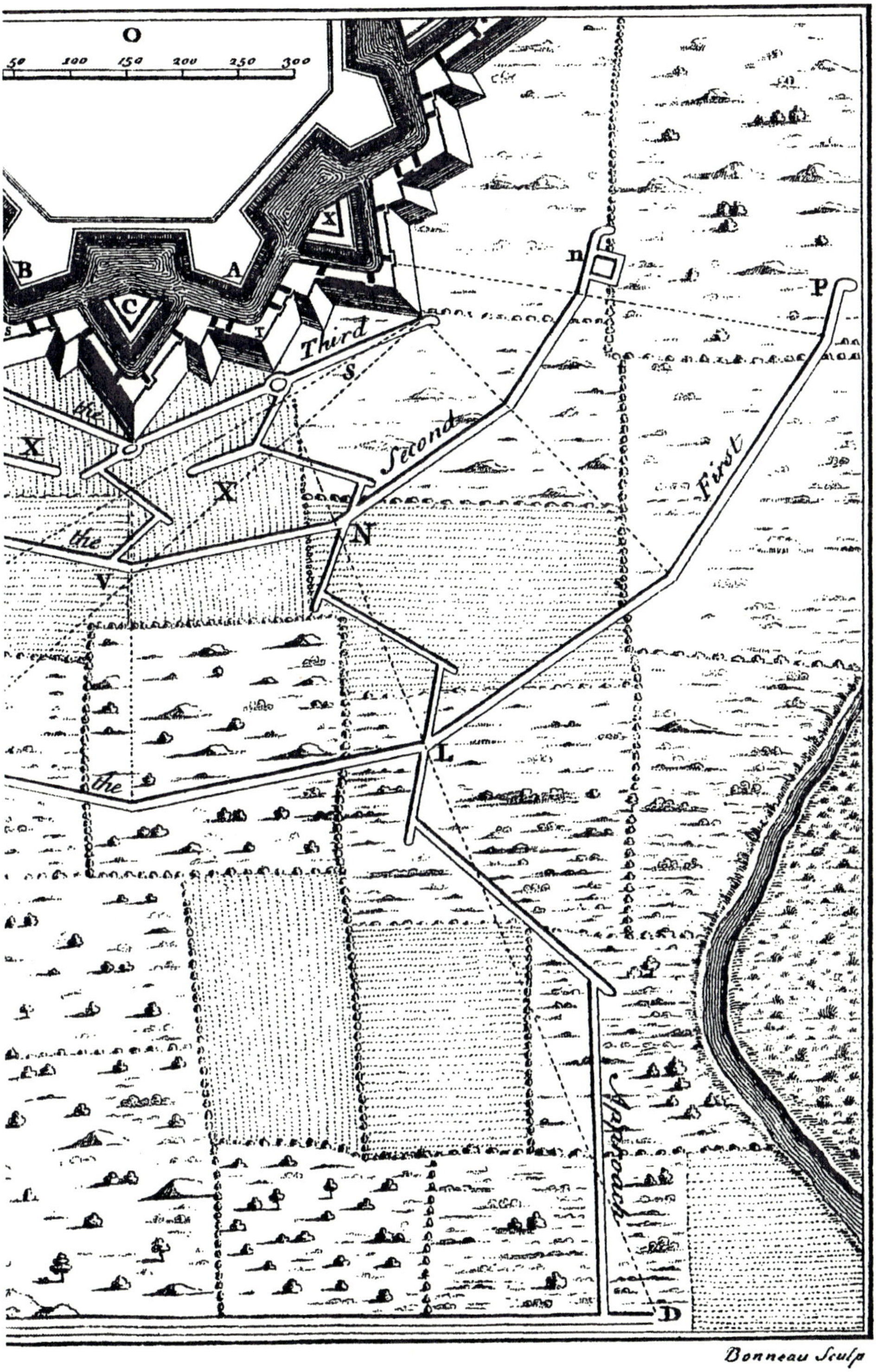

Bonneau Sculp

ſide towards the field, at 12 feet diſtance in the Approaches, 15 in the firſt and ſecond Parallels, and at 18 in the third; then the intervals between theſe and the former lines, will expreſs the width of the Trenches.

At the ends of the ſecond Parallel are made ſquare Redouts, ſuch as m, n, of about 10 or 12 fathoms each ſide, to protect it more effectually from the ſallies of the Beſieged.

The firſt part D E of the Trenches, is an *Epaulement* of about 10 feet high, to cover the horſe deſigned to guard the Trenches; but if there ſhould be ſome riſing ground thereabout, that Epaulement need not be made.

Note, the ends of the Approaches next to the field are produced 2 or 3 fathoms, in order to cover the parts which they terminate, and ſerve for ſome to retire into, upon occaſion, that others may paſs and repaſs, without trouble or confuſion.

The Trenches are commonly 3 feet deep, and the height of the parapet 7, from the bottom of the Trenches. As to the reſt of the dimenſions, they may be ſeen in the profils of the 5th Plate; where it may be obſerved, that the breaſtwork of the third Parallel is made with ſeveral ſteps, ſo that the troops placed there may paſs over it without confuſion, when occaſion requires.

PL. V.

OBSERVATION.

THIS method of tracing the Trenches was invented by M. *Vauban,* and has been in uſe ever ſince, as being undoubtedly the beſt hitherto contrived; yet in my opinion, if the Approaches were carried on in right lines, from the beginning

to

to the firſt Parallel, meeting it nearly in the ſame point as the capital; that is, if the line I H was produced backwards, till it meets the line D E, the Approaches made in that line would be better than the preſent ones, as they would be ſhorter, and require leſs work, and the paſſing and repaſſing attended with leſs confuſion. The tracing them on the ground likewiſe might be done with greater eaſe and exactneſs, as requiring but two pickets to direct them, which might be placed either in day-time, or in the dusk of the evening; and as they are pretty far from the place, the Beſieged would find it a hard matter to enfilade them, which is the only objection that can be made againſt that way of making the Approaches.

It is farther to be obſerved, that the earth dug out of the Trenches is always thrown on the ſide next to the place to make a parapet of it; the ſide where the parapet ſhould be, is expreſſed by a ſtrong line. See Plate IV.

The foregoing method being well underſtood, there will remain no difficulty in the tracing the Attacks, by having a good Plan of the place, tho' never ſo irregular, both in reſpect to the works and the ground about it; but it is not ſo eaſy to transfer the works from the Plan to the ground, eſpecially when it is ſuch as all the parts of the Trenches cannot be equally extended. As the great variety of the ſituations which are to be met with, and which cannot poſſibly be deſcribed, no more than the various manner of executing the Plans can be ſhewn, I ſhall explain how it is done in ſeveral caſes, in order to render it as intelligible to the reader as the nature of the ſubject will admit of.

The

The capitals of the ſeveral works before the front of the attack, are to be marked on the ground, by obſerving the directions of their ſaliant-angles, and the oppoſite ones of the covert-way; and a picket being placed therein as near the Town as poſſible; then going backwards, ſo as to place another in the ſame direction; theſe two pickets will ſerve to place as many more in that direction as are thought neceſſary. Having the capitals on the ground, the points D, E, where the Approaches begin, muſt be found exactly, either by trigonometry or otherwiſe; then meaſuring the diſtance E F on the paper, and ſetting it off on the ground, the point E, where the approaches begin will be found.

PL. IV.

To find the next point G on the ground, draw, in the Plan, a perpendicular from that point to the capital, which being meaſured, as well as its diſtance from the point E, and traced upon the ground, will give the point G; from which a line may be traced with a ſmall rope to the point E. The tracing may be thus continued at pleaſure. It muſt be obſerved, that the perpendicular may be marked on the ground, very nearly by gueſs only; or ropes may be tied together, ſo as to make a right-angled triangle, by which it may be found pretty exact.

The following method is often uſed for the ſake of greater expedition. The Engineers obſerve the directions of the Approaches ſome days before the opening of the Trenches, meaſuring their length by paces, and plant pickets at their returns, in the dusk of the evening; then the next day they obſerve whether theſe pickets are rightly placed, if not, they alter them, and continue ſo in this manner

ner, till they have marked all the turnings as far as the ſecond Parallel; and when the Trenches are to be opened, they lay only faſcines in thoſe directions obſerved before; by this means the works are carried on with eaſe and exactneſs.

If the nights are dark, burning matches or bundles of white ſtraw, are tied to the pickets, in order to be diſcoverable at ſome diſtance.

When the works are advanced pretty near the Town, it would be too dangerous to trace them uncovered; but as they are then carried on by ſaps, the directions may be found by means of inſtruments, and they may afterwards be carried on to the lengths expreſſed in the Plan.

It will be proper to find now and then the diſtances of the works from the covert-way, ſo that if any miſtake ſhould have happened, they may be rectified. The beſt inſtrument for this purpoſe is Dr. *Hadley*'s *Sextant,* by which the angles made by lines drawn from any place, to the ſaliant-angles of the oppoſite baſtions, and to one of the ſhoulders, and the diſtances between theſe points being known from the conſtruction of the place, the diſtance required may be found by trigonometry. See my *Elem. Mathem.* p. 206.

Theſe and other ſuch like methods may be obſerved, in laying down the works from the Plan upon the ground; the ſituation of the place and the nature of the ground about it, will ſuggeſt many other means to proceed by, which can no ways be expreſſed, or thought on before hand.

OBSER-

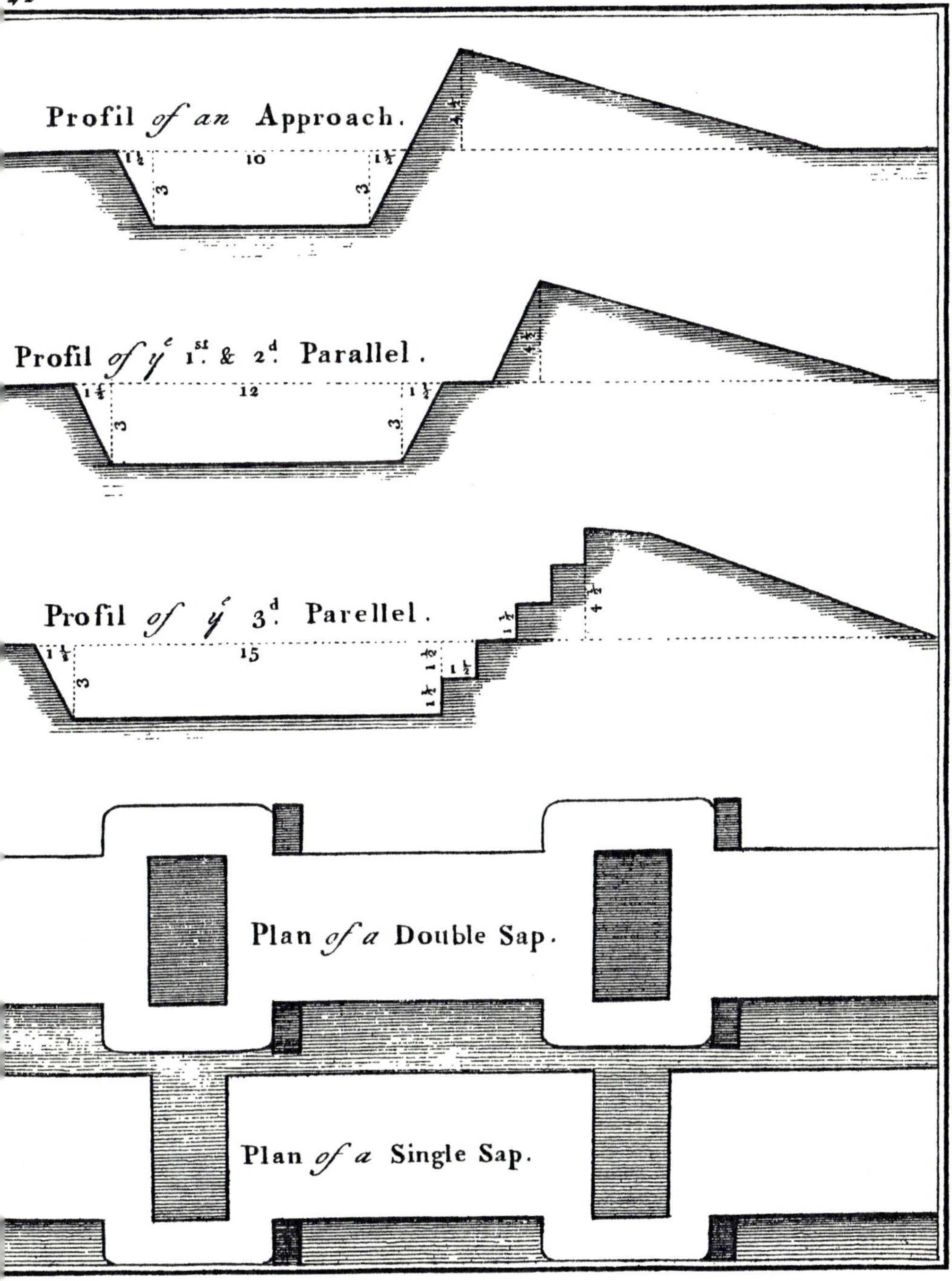
Profil of an Approach.
1½
10
1½
3
3
4½
Profil of ye 1st & 2d Parallel.
1½
12
1½
3
3
4½
Profil of ye 3d Parellel.
1½
15
3
1½
1½
1½
1½
4½
Plan of a Double Sap.
Plan of a Single Sap.

OBSERVATIONS *relating to the properest Place for making the* ATTACS.

HAving a ſufficient knowledge of the ſeveral parts of the Fortification, by means of ſpies, deſerters, priſoners, and from printed or drawn Plans, the nature of the ground about the place muſt be well examined and obſerved, whether there are any hollow ways, or cavins, by means of which the Trenches may be opened nearer than uſual; whether the ſoil be eaſily dug, or if it be gravel, rock, or ſand, marſhy, dry, or wet; whether any part of it may be laid under water, by means of dykes or ſluices, or the moraſs may be drained, or paſſed over with ſome trouble.

If a Town be ſituated near a river, with a Fort on the other ſide, it will be beſt to attac that ſide next to the river, ſo that one flank of the Attacs being ſecure from any attempt that way, the troops may be at liberty to act with greater force on the other. Oppoſite to this Attac, on the other ſide of the river, near the Fort, muſt be another Attac made, and Batteries ſo placed as to deſtroy the bridge of communication between the Town and the Fort; and when the Fort is taken, batteries are to be raiſed near, or in it, to flank and enfilade the works of the front attacked.

But if the river paſſes through the Town, and is pretty large; then the Attac ſhould be made upon the front fartheſt from the river, otherwiſe the enemy may raiſe batteries on the other ſide of the river, to enfilade and annoy the Trenches; and if to prevent this, another Attac is made on the other ſide, they muſt be both equally ſupported, leſt,

as

as they cannot aſſiſt one another in caſe of need, the Beſieged ſhould fall upon one of them, with their whole ſtrength, and deſtroy it.

If there be an Iſland near the Town, from whence any of the works may be ſeen or enfiladed, it will be neceſſary to get poſſeſſion of it, and erect batteries there. If there happens to be any marſhes, which cannot be drained or paſſed over, it will be proper to flank one ſide of the Attac with them, for the ſame reaſons as when near a river; but if the marſhes are to be croſſed by any means whatſoever *, unperceivable by the Beſieged, the doing ſo will accelerate the Siege pretty much; becauſe places are always leaſt fortified on thoſe ſides.

The place and ground about it having been ſufficiently examined, the chief Director of the Engineers makes a Plan of the Attacs, and ſpecifies in writing the particular reaſons why he chuſes to attac the place on that ſide, which he gives to the General to have his approbation.

The Commander in chief of the Artillery is alſo to be conſulted about the number and diſpoſition of the batteries, as being, or ſuppoſed to be the beſt Judge of what belongs to that branch of the buſineſs.

It was formerly the cuſtom to make ſham Attacs, only to amuſe the Beſieged and conceal the real one; but as they are found out in about three or four days, and are an additional expence and labour, they have been neglected of late.

It

* At the Siege of *Bouchain*, the late General *Armſtrong* carried a work through marſhes full of reeds, in one night's time, by means of hurdles and faſcines; which contributed much to the ſpeedy ſurrender of that Place.

It is true, that when the place is of great importance, and the Beſiegers Army numerous, they generally make more than one Attac, eſpecially when there is a Citadel; then they uſually carry on one againſt it, and another againſt the Town, as the *French* did in 1745 at *Tournay*.

The Attacs ſhould never be ſeparated, unleſs the Town is weakly garriſoned, or the Beſiegers very numerous. The double ones are much preferable, as they may afford mutual aſſiſtance to each other, and fatigue the garriſon as much as if they were ſeparated; and no ſingle Attac ſhould ever be made, but when there is not ſpace enough to make two, as it ſometimes happens when the place is ſurrounded with marſhes, and there is no way of coming at it, but by cauſeways. Then, the utmoſt precaution ſhould be uſed to ſupport every part of it, in the beſt manner poſſible, by making places of arms, from diſtance to diſtance, with faſcines and gabions, for want of earth, to prevent the Beſieged from falling on the workmen and their guard, and deſtroying the works.

The Plan of the Attacs being ſettled, the little parks of Artillery, as alſo the hoſpitals, are to be placed as near the opening of the Trenches as poſſible, for the more eaſy carriage of the Artillery, ammunition, and ſtores, into the Trenches, and bringing the ſick and wounded out of them.

Theſe Parks are to contain a ſufficient quantity of artillery, ammunition, powder, and ſtores, and all kinds of tools for digging the Trenches, as alſo for ſapping; in ſhort, every thing neceſſary for the daily uſe of the Trenches; and as faſt as they are either expended or ſpoiled, they muſt be ſupplied from the great Park. The little hoſpitals are made

made of tents only, as being ſuppoſed out of the reach of cannon-ſhot from the place; they ſhould be provided with chaplins and ſurgeons, all kinds of medicines and apparels for the immediate uſe of the ſick and wounded brought from the Trenches.

The place of Parade, where the troops ordered for the Trenches are drawn up, is alſo to be as near the openings of the Trenches as can be, that they may not have a great way to march; for the ſame reaſon the General's quarters ſhould be likewiſe near at hand, as has been obſerved, to ſee and order every thing required for the daily ſervice.

The roads from the ſeveral quarters, to the rear of the Attacs, ſhould be made on the drieſt ground, out of the reach of cannon-ſhot; care muſt be taken to keep them in conſtant repair, ſo as to be at all times an eaſy ſafe paſſage.

If a ſpring or brook ſhould happen to be near the rear of the Attacs, care ſhould be taken that the water be kept clear, and no dirt or naſtineſs thrown in, being one of the moſt uſeful things for the ſupport of mankind; for which reaſon a Guard ſhould be placed over it, and if the guns of the Beſieged ſhould be able to interrupt acceſs, or make it dangerous, a trench muſt be made to it.

The Engineers quarter, as likewiſe that of the miners and ſappers, ſhould alſo be near at hand, as being the moſt neceſſary people in carrying on the Trenches; and therefore the nearer they are, the better.

The Attacs being reſolved upon, the guard for the Trenches ſhould be regulated in ſuch a manner, as to be at leaſt equal in foot to three fourths of the Garriſon, and to exceed it by one third more in horſe; ſo that if the Garriſon conſiſts of four thou-

thouſand foot and four hundred horſe, the Guard is to be three thouſand foot and ſix hundred horſe.

It was formerly computed, that the Army of the Beſiegers ſhould be ten times the ſtrength of the Garriſon, but this rule can only be true in ſome particular caſes; for when the Garriſon is either very ſmall or very large, the Army would be too ſmall or too great. For example, if the Garriſon conſiſts of a thouſand men, then, according to this rule, the Beſiegers ſhould be ten thouſand, which are too few, ſince the leaſt Town requires a Line of circumvallation of ten miles; ſo there would be but one thouſand men to guard a mile of the Line, and furniſh troops beſides for the Guard and works of the Trenches; which would not be ſufficient to prevent the enemy's throwing ſuccours into the place; on the other hand, if the Garriſon were ſixteen thouſand men ſtrong, as that of *Liſle* was, when beſieged in Queen *Anne*'s War; the Beſiegers ſhould then be one hundred and ſixty thouſand, which are a great many more than are neceſſary.

In my opinion, there can be given no general rule, whereby the number of troops required for a Siege of all ſorts of places, may be determined, as depending on too many different conſiderations, without the exact knowledge of which there can be no ſuch thing as making a true eſtimate. At preſent, Sieges are carried on with ſuch numerous Armies, and ſo great quantities of Artillery, as that it is hardly poſſible for a Garriſon to make a long reſiſtance; and it is beyond diſpute, that the greater the Army is in compariſon to the Garriſon, the ſooner the place will be obliged to ſurrender.

The number of Engineers neceſſary to carry on the Siege, ſhould be thirty at leaſt for a midling place,

place, and from forty to fifty for a ſtrong one. As they are abſolutely neceſſary in a Siege, there cannot be too many of them, and being more expoſed than any others, there are always many of them killed and diſabled. They are divided into Brigades of ſix men each; viz. a Director, a Subdirector, and four Engineers.

As to the number of Artillery men, there ſhould be in my opinion, no leſs than a battalion of eight hundred including the miners and ſappers. The *French* have never leſs than two compleat battalions.

Of Opening the TRENCHES.

WHEN every thing is ready, the Line of circumvallation being nearly finiſhed, and a great quantity of faſcines and pickets at the front of the camp, and all kind of neceſſary tools in the little parks, as likewiſe the Artillery in readineſs, the day for opening the Trenches is reſolved upon. The General ſettles the Tour of duty for the guard of the Trenches both of horſe and foot, ſo as to have five or ſix days reſt; alſo the number of horſe for bringing the faſcines and pickets from the front of the camp, to the place of opening the Trenches, and the number of pioneers for each day, and each night: All which is done a day or two before the Trenches are opened.

The Major-Generals of the horſe and foot conſult with the chief Director about the number of men neceſſary for each day's ſervice, and take care to have them in readineſs whenever they are called upon.

The chief Director ſettles alſo, with the reſt of the Engineers, the particulars of their buſineſs, and gives

gives them copies of the Plan of the Attacs, with written inſtructions, ſo that they may the better follow his orders; and they take care to be provided in time with pickets, mallets, rods, and ropes for tracing the Trenches.

A ſmall guard is placed at the rear of the Attac, to take care of the tools and inſtruments lodged there, and not to ſuffer the place to be too much frequented, it being prudent to conceal the deſign from the enemy, as long as poſſible.

The day for opening the Trenches being come, the guards parade about two or three in the afternoon, and prayers are ſaid; then they file off before the General, if he think proper. The pioneers meet likewiſe hard by, having each a faſcine, picket, ſpade, and a pick-ax.

In the dusk of the evening the troops advance; the grenadiers, and other detached bodies, who are to protect the workmen the firſt night leading the van, followed by the battalions deſigned for the guard of the Trenches, and the pioneers in the rear; theſe laſt are, according to M. *Vauban*, divided into brigades of fifty men, each commanded by a captain, a lieutenant, and two ſerjeants, and march four or ſix in front. Each ſoldier of the guard carries a faſcine beſides his arms every time they mount during the whole Siege.

When the front of the troops is arrived at the place where the Trenches are to be opened, the Director of the day leads the brigades of the pioneers forward, and places them where the Trenches are to be made, whilſt the guards of foot place their faſcines at the opening, retire behind ſome cover, if there is any, or where the Major orders them, and lie down upon their arms, ready

ready for ſervice whenever they are called upon. The guard of horſe march likewiſe to their aſſigned poſts, at the right and left of the opening, ready to ſupport the foot, in caſe of any ſally from the enemy.

In the mean time, the Director places the firſt rope, and ſhews to the reſt of the Engineers what is to be done, that they may continue the work; then he makes the pioneers file off one by one, each carrying his faſcine under the right arm, if the Town is on the right, or under the left, if it is on the left, and begins himſelf to place the firſt pioneer, then the ſecond, third, fourth, &c. ſucceſſively one after another, ordering ſilence, all to lie down, and not begin to work till commanded.

When the Director has placed ſeveral pioneers, he leaves the firſt Engineer to place the reſt, whilſt he goes to ſee how the tracing is carried on; he takes care that the branches of the approaches next to the field, are produced two or three fathoms, as was ſaid before, to cover thoſe which they terminate, from being enfiladed; that the faſcines are laid on the ſide next the Town, and that the Approaches, when produced, fall neither too far off nor too near the fartheſt ſaliant angles of the covert-way, but within about ten or twelve fathoms of them.

It is ſuppoſed that the chief Director has taken ſuch meaſures as to carry the Trenches ſo far as the firſt Parallel, the firſt night, and that even a part, if not all, of that Parallel, be made.

Whatever has been ſaid with regard to one Attac, ought likewiſe to be underſtood with regard to the other; and it is to be obſerved, that they ſhould

ſhould both advance together equally; and if one ſhould meet with any obſtacle, the other muſt wait for it; to prevent the enemy from taking any advantage of that which is moſt advanced.

When every thing is ready, the word or ſignal is given to begin the work, taking care that the earth is thrown on the ſide directed, and all poſſible diligence is to be uſed till day appears; then part of the detachment retire behind the Parallel began, and the reſt behind the head of the Approaches, and lie down on their faces, the works being yet imperfect.

The workmen of the night are relieved in the morning by an equal number who begin to work at the head, contrary to thoſe of the night, who began at the rear; theſe are to finiſh, if poſſible, what the others have begun; but if that cannot be done, a detachment of 100 or 200 men are ordered extraordinary the ſecond day to finiſh it.

As the deſign is known to the enemy, the guard mounts the next day, with drum beating about four in the afternoon, in order that the commander may have time to viſit the works, and know what is to be done next night, in which the workmen advance uncovered in the ſame manner as the firſt, and they continue ſo to the ſecond Parallel, always finiſhing by day what was begun by night; and as ſoon as the works are in a condition to receive troops, they retire into them, with order to thoſe at the head of the Approaches, together with the workmen, to retire alſo, in caſe the enemy ſhould make any ſally; as not being yet in a condition to repulſe them.

The ſecond day as many workmen as are neceſſary are ordered to finiſh what was left to be done, eſpecially efpecially the firft Parallel, being the conftant receptacle of the guards for the Trenches, till the fecond is quite finifhed.

The work is thus continued day and night with all poffible fpeed, advancing as far by night as poffible, and finifhing by day what was begun by night; and if the fame number of workmen is not fufficient, more are ordered for that purpofe, that nothing may be wanting to carry the works with all expedition as far as the fecond Parallel.

When the works are advanced to the fecond Parallel, the fire of the enemy becomes too dangerous to proceed farther uncovered; and therefore it will be proper to carry on the works by faps; tho' in the night, when their fire begins to flacken, fome part may be ftill carried on uncovered; but this fhould be done with great caution; for it is an inviolable maxim never to expofe men without an abfolute neceffity, which however is too frequently done by forward Commanders, who think little of facrificing many brave men's lives to their own hafty Ambition.

Before we proceed farther with regard to the Trenches, it will be neceffary to fhew how the batteries are placed before the firft Parallel, referving to treat of the others in due order.

BATTERIES *placed before the firft* PARALLEL.

THE firft Batteries are no where fo conveniently placed as about fifty toifes before the firft Parallel, and if the ground is any ways advantageous, they may continue there during the whole Siege.

The

The Intent of the batteries will naturally point out their ſituation; they ſerve for two purpoſes, *viz.* to diſmount the Beſieged's guns, and drive the enemy from their defences, by enfilading the faces of thoſe works which command the front of the Attac.

A Battery intended to diſmount the guns of the Beſieged, ſhould be directly or nearly oppoſite to them; and in order to enfilade any work, it ought to be perpendicular to that work.

Hence the Battery E, which is perpendicular to one of the produced faces of the baſtion A, is alſo nearly parallel to the other face, and may therefore either enfilade the face to which it is perpendicular, when fired *à Ricochet*, or diſmount the guns placed upon the other face to which it is nearly parallel. It may be obſerved, that it is not abſolutely neceſſary that the Battery ſhould be directly oppoſite to the face, in order to diſmount the guns placed on it; for by being a little oblique, there is a better chance of a ſhot's hitting one gun if it miſſes another.

PLATE VI.

The Battery F, which is perpendicular to one of the faces of the baſtion B, will enfilade it by firing *à Ricochet*, and diſmount the guns placed on the other face, when fired in the uſual manner. Likewiſe the Batteries G, H, which are both perpendicular to one of the faces of the ravelin C, will enfilade that face to which they are perpendicular, and diſmount the guns placed on the other.

From whence it appears, that as many Batteries are required in an Attac, as there are faces of the works which flank the Trenches: and therefore there are eight Batteries required in ſuch an Attac as this, *viz.* two for each of the baſtions A, B;

two

two for the ravelin C; and one for each of the adjacent ravelins x, z.

It is to be obſerved, that the Battery which enfilades any of the faces which flank the Attac, will alſo enfilade the covert-way parallel to that face. There ſhould be no leſs than four guns on each Battery, becauſe the ſlowneſs of the *ricochet* would give the Beſieged an opportunity of making *Traverſes,* to ſcreen themſelves from that dangerous fire; for which reaſon the fire ſhould be continual.

Beſides the Batteries of cannon, there are alſo Batteries of mortars, in the ſame line with, and joining thoſe of cannon, to throw ſhells into the works.

Thoſe for cannon are made with *Embraſures,* and thoſe for mortars, with an *Epaulement,* or breaſt-work, without *Embraſures:* Trenches are made from each end of the Batteries to the firſt Parallel, or to the next Approaches, to go to, or from the Batteries without danger; and a paſſage is made over the Parallels for carrying the guns and mortars that way to the Batteries, which are afterwards clear'd to preſerve the communication from one part to another. When guns are fired *à Ricochet,* they are loaded with a ſmall quantity of powder, and the gun is elevated about ſix or ſeven degrees above the level, that the ſhot may juſt riſe above the oppoſite parapet, and after alighting bound and roll along the rampart behind the other parapets, for which purpoſe meaſures of ſeveral ſizes are made, to find the proper charges, which once found is not to be alter'd.

M. *Vauban* was the firſt inventor of the firing *à Ricochet* with guns, which has ſince been applied

to

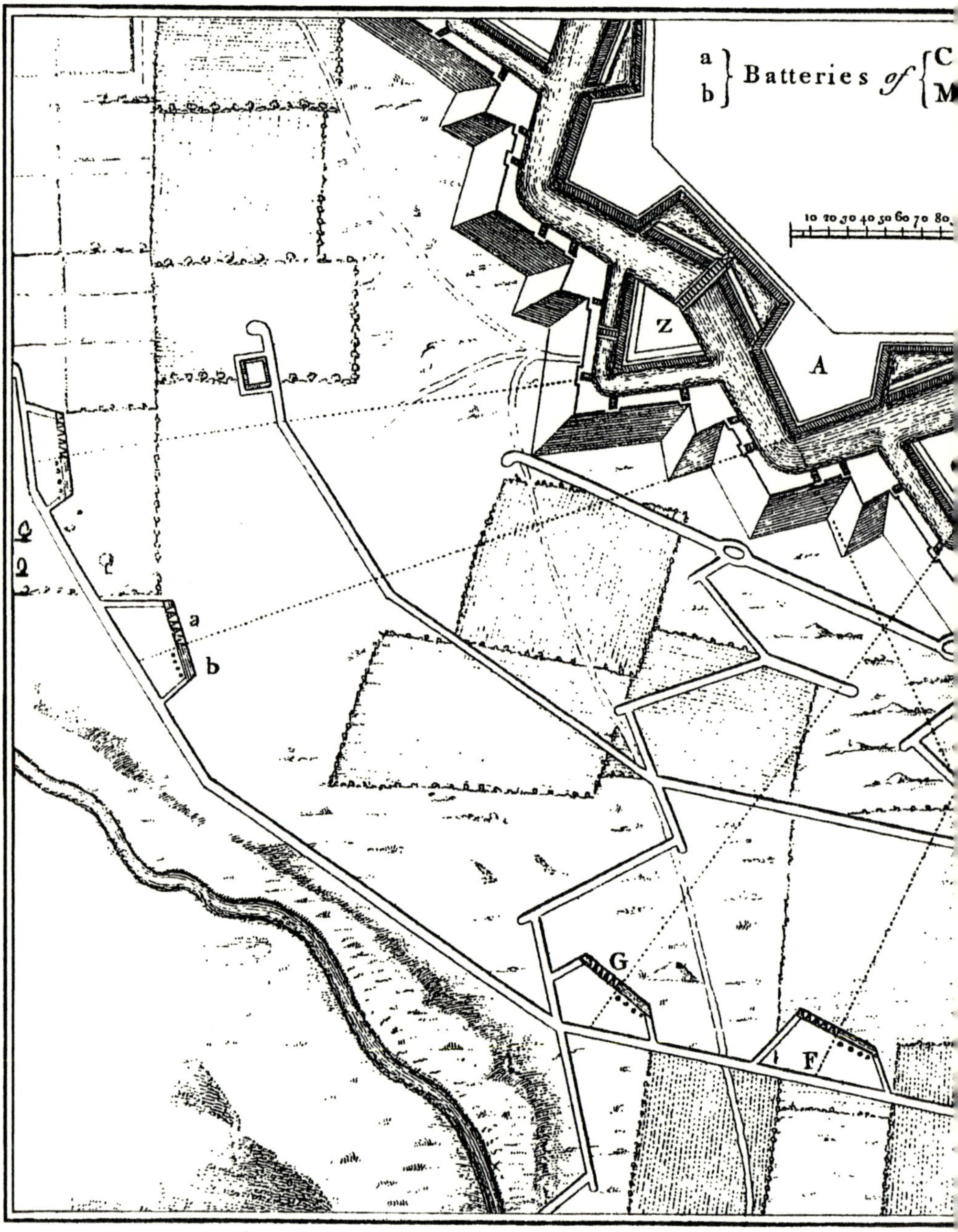
a
b
Batteries of
10 20 30 40 50 60 70 80
Z
A
a
b
G
F

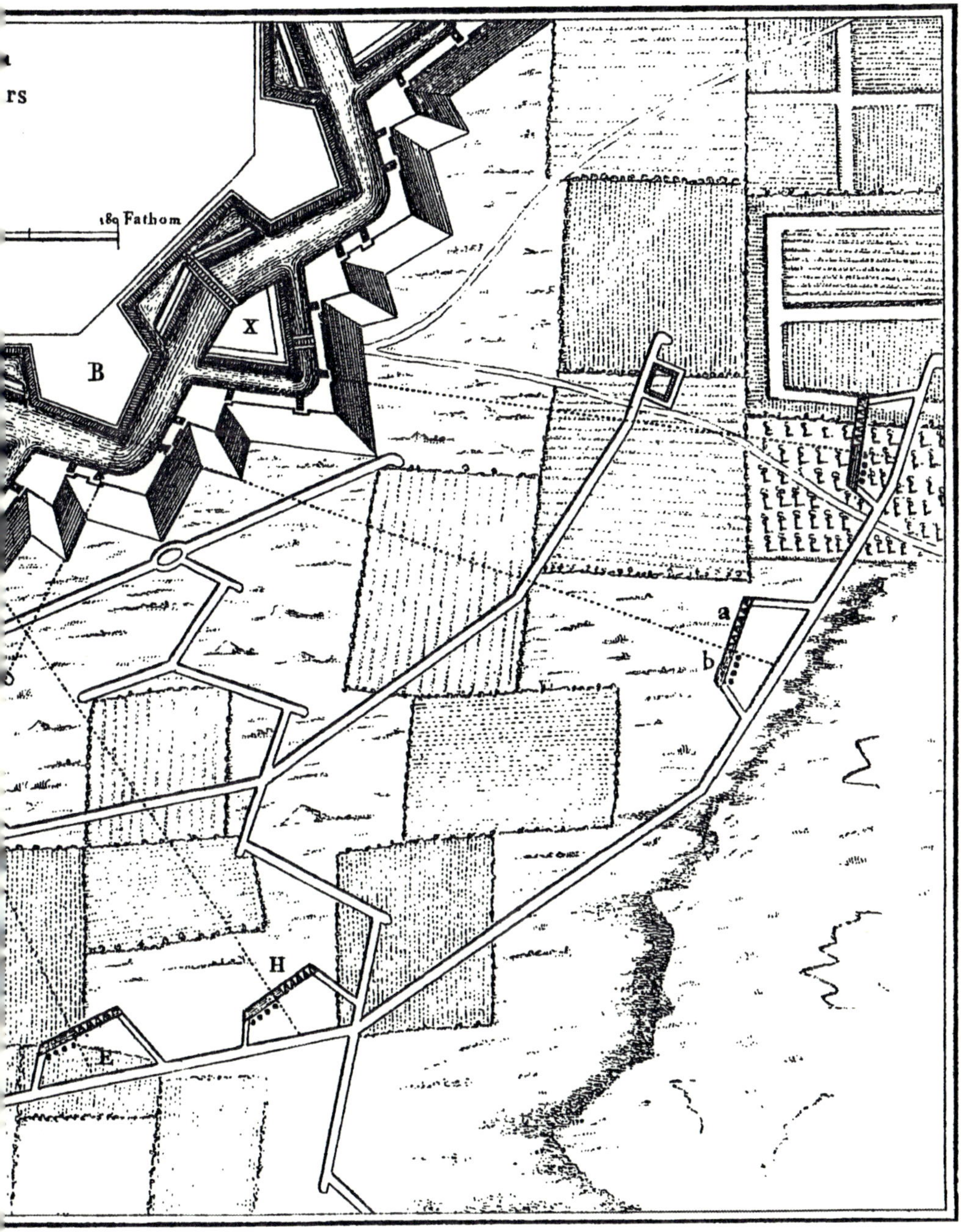
rs
180 Fathom
X
B
a
b
H
E

to mortars and *howitzes* with great ſucceſs: It is certainly the moſt advantagious manner of firing that can be imagined; as it requires but little powder, it does but little damage to the guns, yet incommodes the Beſieged in their works, without making a great number of traverſes, which are very troubleſome in ſtopping the free communication of one part with another, yet not entirely prevent the effect; and when mortars and howitzes are fired in that manner, their ſhells do the execution both of ſhells and ſhot.

CONSTRUCTION *of* BATTERIES.

PLATE VII. THE Conſtruction of Batteries belongs to the officers of artillery, altho' here in *England* the Engineers are employed in theſe works, whoſe Commander in chief ſhould therefore conſult with the Director about their places, which when known, Trenches are carried from the Parallel, or Approaches, quite round them *; then a detachment of pioneers take the earth from the ſide of thoſe Trenches, next to the Town, and throw it backwards, to raiſe a parapet,

* The *French* have a great number of Artillery Officers called Commiſſaries, independant of the regiment; who are employed in the managing and placing ſtores in the park; to make and command the Batteries. The King pays commonly 300 livres for every gun placed on the firſt Batteries, 500 for every piece placed on the covert-way, and a 1000 for each placed in a ravelin, or counter-guard, or any other outwork; he pays beſides 10 or 15 livres a day for each piece while fired, by way of ſubſiſtence money: Out of theſe ſums the expences of making the Batteries, which are very conſiderable, in regard to the wood for making plat-forms, faſcines, and paying the workmen. He pays 500 livres for each mortar; 16 livres a day for ſubſiſtence money, and but 10 for pieces fired en barbet. What remains at the end of the Siege, is divided amongſt the officers employed in proportion to their rank (the commander excepted) who receives a gratification from the Maſter General.

pet, or breaſt-work: In the mean time the gunners * prepare the materials for the platforms. The breaſt-work ſhould be 18 or 20 feet thick, and ſeven and a half high, lined with faſcines made for that purpoſe, well faſtened together with pickets or ſtakes.

There are generally 200 pioneers and 20 gunners at making a Battery of 10 guns, and for larger or ſmaller in proportion. As ſoon as a ſufficient quantity of earth is thrown up to cover the gunners within, they make a ſmall Trench of five or ſix Inches deep from one end to the other, in which they lay the firſt faſcine, and fix it well to the ground with pickets; the next is laid at the end of it, and is fixed to the ground in the ſame manner, and ſo continues to the end of the Battery. Then they begin another row at the top of the firſt, faſtened to it with pickets, taking care that the ends of the upper faſcines do not correſpond with thoſe underneath them, for which reaſon ſome faſcines are made ſhorter than others. When the ſecond row is finiſhed, they lay a third, and continue in this manner the work quite up to the top, obſerving to give it a little ſlope, that the faſcines may better ſupport the earth, and to turn the ends of the bindings inward, which gives a neatneſs to the work.

When the faſcines are raiſed three feet above ground, there muſt either be left openings of two feet wide at proper diſtances for the embraſures, or the work may be carried up entire to the top, and the embraſures opened afterwards, by cutting the faſcines with ſaws.

If

* In *England* theſe Works are done by Engineers.

P: 39

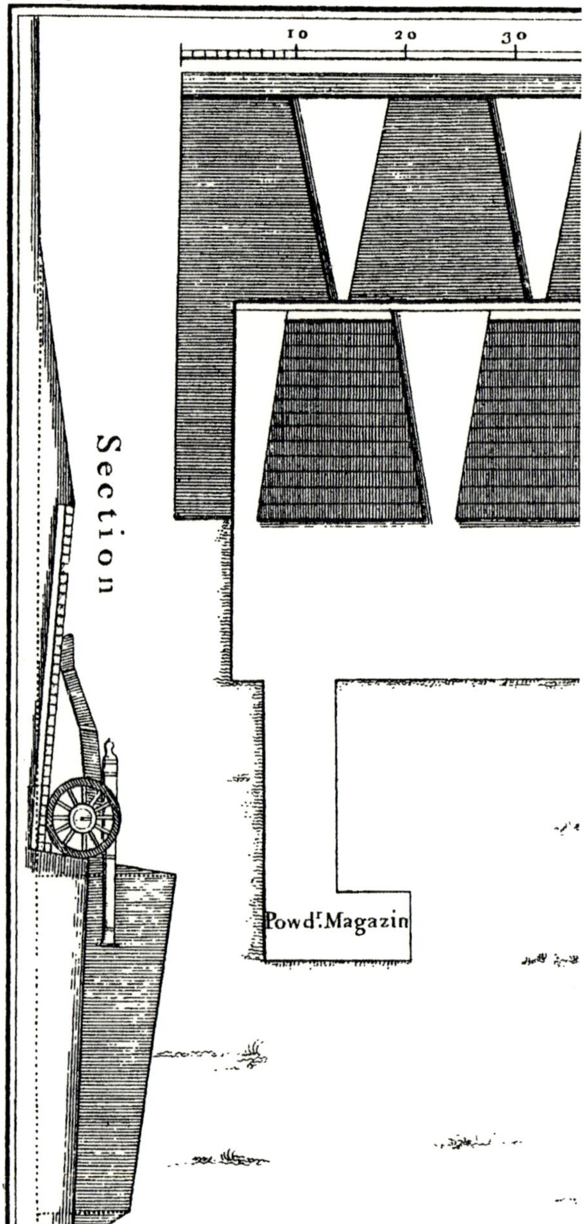

VII.

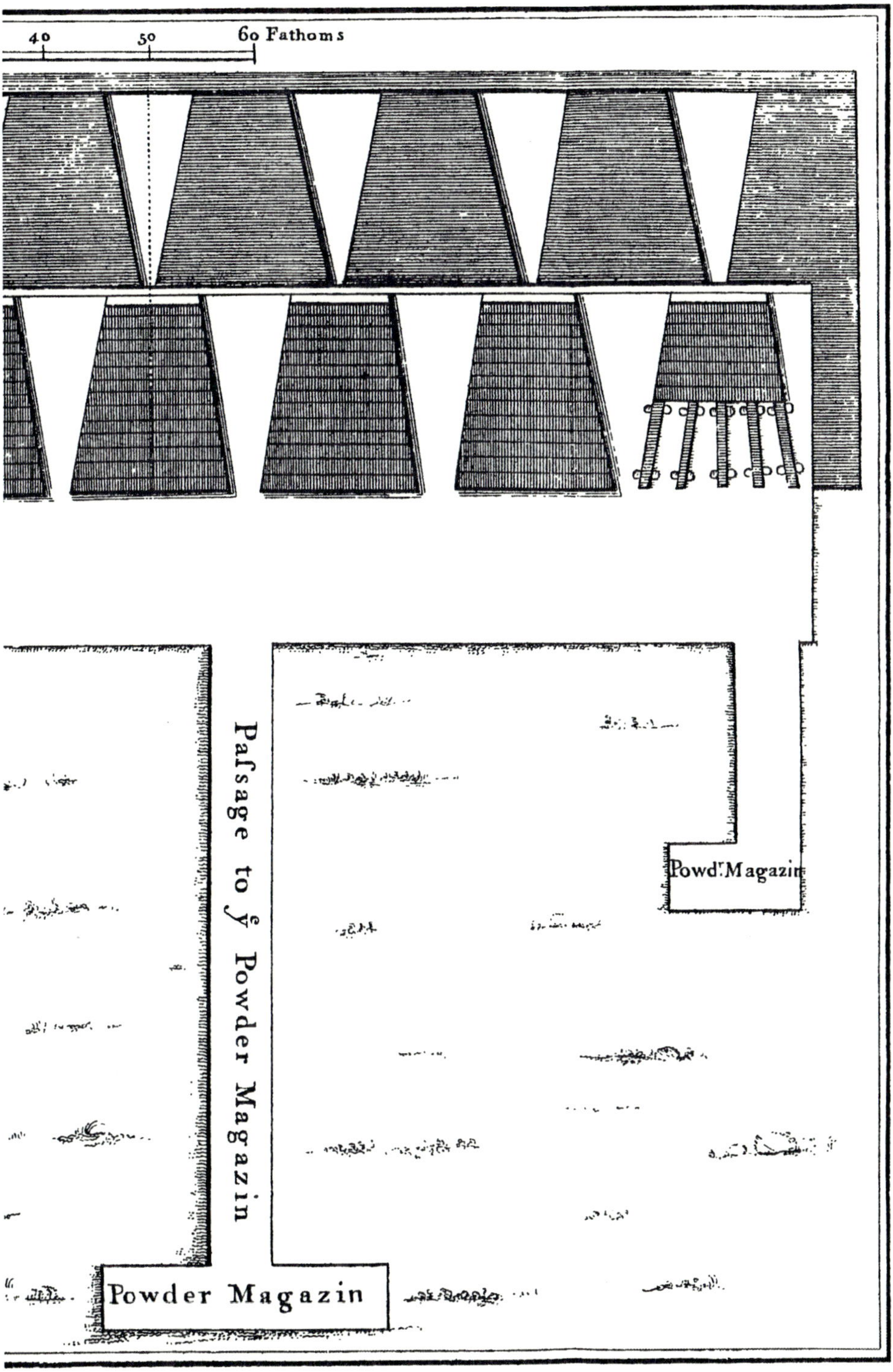
40
50
60 Fathoms
Paſsage to ye Powder Magazin
Powdr. Magazin
Powder Magazin

If the Battery happens to be flanked by any of the works of the town, the flanks muſt be cover'd by a parapet of 10 or 12 feet thick, and made in the ſame manner, and at the ſame time as the breaſt-work.

The diſtance from the center of one embraſure to that of the next, is generally three fathoms or 18 feet; ſo that a Battery of 10 pieces of cannon is 30 fathoms long on the inſide; the embraſures are two feet wide within, and about nine without, ſlanting outwards about a foot and a half.

Whilſt the earth is throwing up for the parapet, the gunners ſhould lay the platforms, as ſoon as there is enough thrown up to cover them, beginning to lay five Joiſts or Sleepers longways from the parapet, ſecuring them on both ſides with ſtakes; then the Hurter is laid next to the parapet, which is a piece of timber about Six inches one way, and five the other; and after that the planks of about three inches thick. The earth muſt be well ramm'd betwixt the joiſts, and round the edge of the platform, to make it ſtrong and firm. The platforms are 15 feet broad behind, 9 before, and 18 long, with a ſlope forwards, of about nine or ten inches.

When there is a ſufficient quantity of earth thrown up for the parapet, and ſmoothed, the gunners begin to open the embraſures from within, leaving always ſome earth before them as long as they can, to cover them againſt the fire of the place.

They begin to plant two pickets, one within and the other without, to mark the center, or middle line of the embraſure, and then plant two more on each ſide, to mark the openings within and without:

without: This done, the earth within the embrafures is removed for the fpace of 10 or 12 feet, and fafcines laid, well picketed, to fecure the fides of the embrafures; and when this is done, they make trenches on both fides, juft wide enough to lay the fafcines, and leaving the fpace of three or four feet in the middle, to cover them; and when thefe fafcines are laid, and the guns ready to fire, the remaining earth in the embrafures is taken away, which is called *Unmasking* the Battery.

A ftuft gabion is laid acrofs the embrafure on the infide of the parapet, to keep off the musket-fhot; and the gunners generally make ufe of a piece of wood three inches thick, juft fitting the infide of the embrafure, for the fame purpofe, whilft they are loading or cooling the piece.

They alfo make ufe of an *Aim Frontlet,* which is placed upon the vent field of the gun, when they point it, and which is a piece of a plank three inches thick, a foot long, and 7 or 8 inches high, with a round cavity underneath, to fit the outfide of the gun, and having a fmall flit, to fee the object through it.

About 20 or 30 yards from the Battery, towards the parallels, a Magazine is made, about three feet under ground, to hold the powder and loaded fhells. Its fides and roof are well fecured with boards, to prevent the earth from falling in; a door is made to it, and a double trench or paffage is funk from the magazine to the Battery, one to go in, and the other to come out, to prevent confufion.

If the paffages leading to the magazines are pretty long, there might be traverfes made in them to prevent the plunging into them by the *Ricochet* batteries,

batteries, or they may be made from the ends of the Battery, and ſo turn in a curve line.

If the Battery is more than of four cannons or mortars, there are generally two or three magazines made, the middle one being much larger than the others, and farther from the Battery. This large one is, as it were, the great or general magazine, from which the leſſer are ſupplied upon occaſion; it is prudent to have more than one, ſo as not to have too much powder lodged together, leſt any accident ſhould happen.

It were to be wiſh'd, that the Batteries of cannon were raiſed five or ſix feet above the level ground, to plunge the better into the works; but when the ground is not naturally high, it would be too much trouble to raiſe it; for which reaſon they are only made even with the level ground.

As to the mortar Batteries, they differ from the former, only in having no embraſures, and that the ſlope of the breaſt-work is made inwards contrary to that of other parapets, their platforms are but ſix feet ſquare and eight feet aſunder. Theſe Batteries may be ſunk in the ground, and the earth taken out ſerves to raiſe the parapet, which will ſhorten the work; mortars may alſo be placed in the Parallels.

PLATE VIII

The bombardiers plant two pickets or ſtakes on the parapet in the direction of the object, which ſerve them to lay the mortars by, as they have no ſight of the object itſelf.

In *England* the mortar-beds are made ſo very large and heavy, that no platforms are required; but whether it is the beſt way or not, we ſhall leave it to the judgment of the Commanders of Artillery. It is certain where the ground is hard and firm,

firm, the Battery is ſooner made; but the beds and mortars together, being very heavy, and not eaſily removed from one place to another, and require much trouble to manage them; and when the ground is ſoft, the ſhock of the mortar firing, makes the direction of the ſhell very uncertain.

Of the SAP.

I Have obſerved, that when the Approaches are advanced ſo far as the ſecond Parallel, the fire of the Beſieged becomes too dangerous to go uncovered, tho' it is ſuppoſed, that all or moſt of the guns are diſmounted by that time; otherwiſe it would be impoſſible to proceed farther, for which reaſon I ſhall here ſhew how to proceed under cover, tho' with ſomewhat leſs ſpeed, yet with greater ſecurity, that when uncovered.

The advantage of this work is, that it may be continued day and night, ſince the workmen are covered on the flank by gabions filled with earth, and faſcines over them, and on the front by mantlets or ſtuft gabions, which they roll before them as they advance. PLATE IX.

The ſappers are divided into brigades of eight men each, to relieve one another, and theſe brigades are ſubdivided into four and four; ſo that whilſt four of them carry on the work, the other four furniſh them with the neceſſary materials; and when the firſt four are tired, the others relieve them, the firſt furniſhing thoſe in their turn with materials.

The Engineers ſhews the firſt ſapper where to make an opening in the Trenches, and gives him directions; then he begins by making room

for

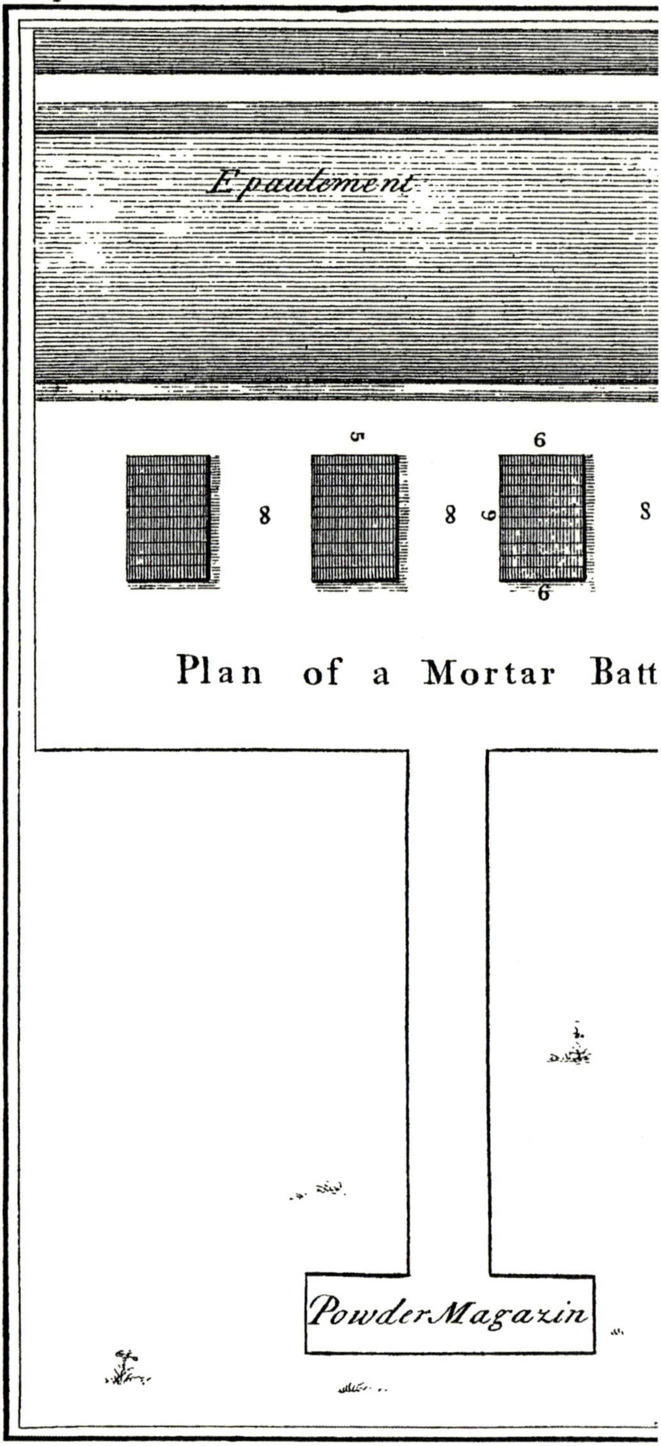
Epaulement
5
6
8
8
6
S
6
Plan of a Mortar Batt
Powder Magazin

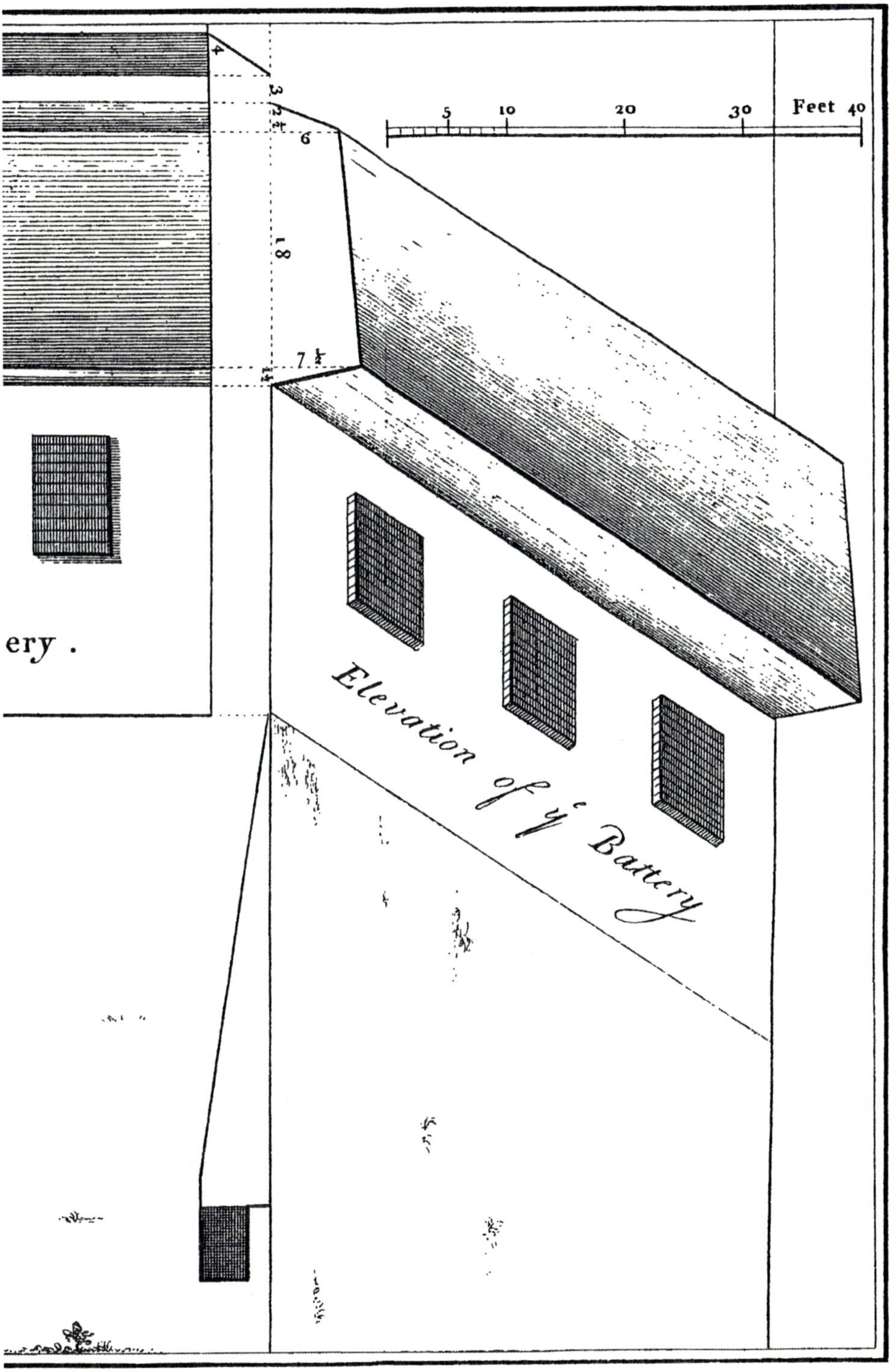
5
10
20
30
Feet 40
4
3
6
18
ery .
Elevation of yͤ Battery

P : 44

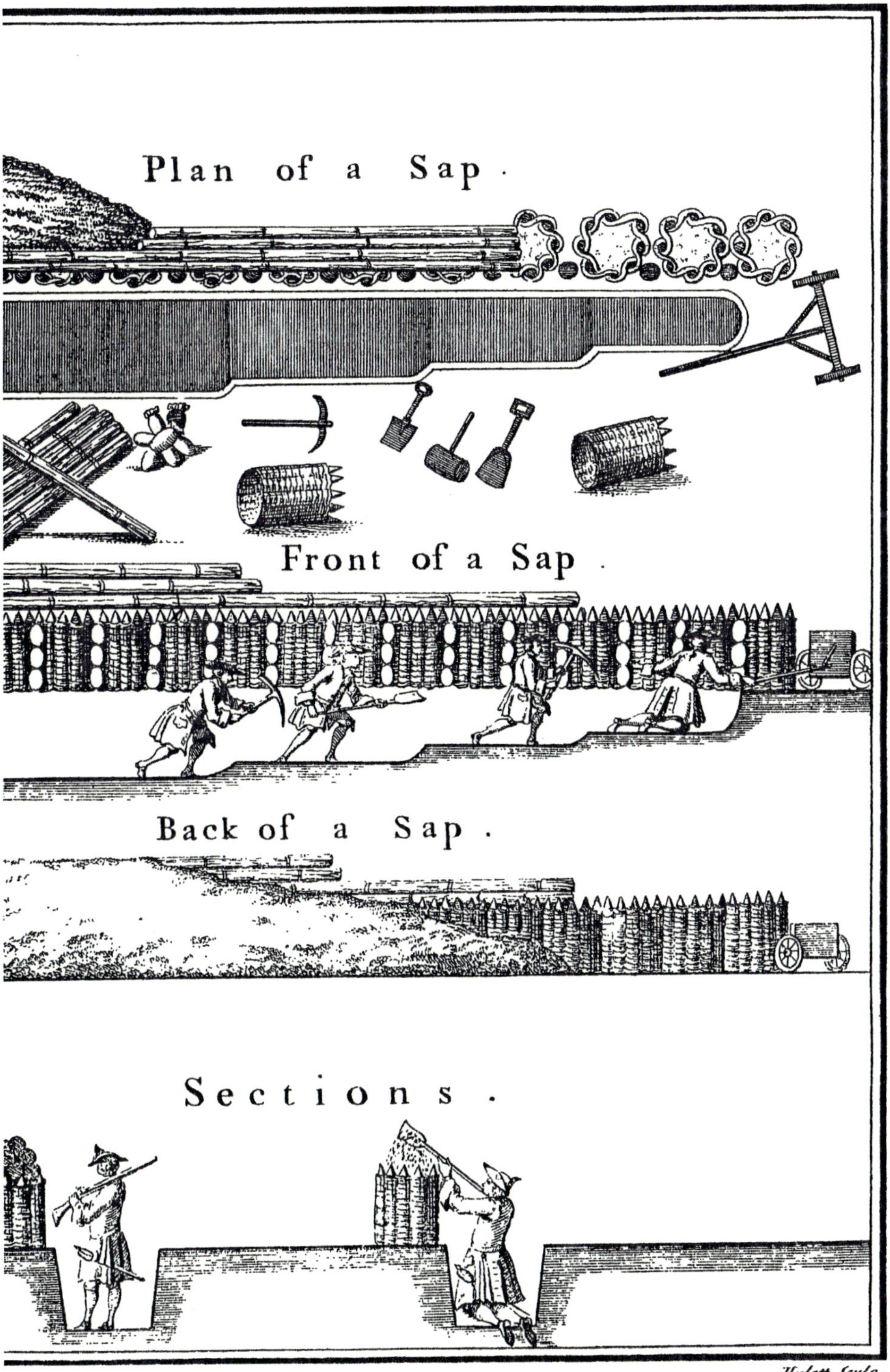

Hulett Sculp.

for the mantlet or gabion, and ſets the firſt gabion with the points upwards to place faſcines on them afterwards; he then fills it with earth, giving it every now and then a blow with the ſpade or mallet to ſettle the earth; when this gabion is filled, he advances the mantlet or ſtuft gabion, to make room for another, which he places by means of the iron hook and fork deſigned for that purpoſe; and this he may do without expoſing himſelf. When this gabion is filled, he places another, and ſo on; then the ſecond ſapper fills the interval betwixt the gabions with ſand bags fill'd with earth, or with ſap faggots.

The firſt ſapper makes a Trench of 18 inches deep and as wide, the ſecond makes it ſix inches, and widens it as much; the third makes it ſix inches more, and widens it as much; and the fourth does the ſame; ſo that all four together make a trench of three feet wide, and as much in depth; the earth of which will ſerve to make a parapet ſtrong enough to reſiſt musket ſhot.

When a ſufficient number of gabions are placed, the third and fourth ſappers lay two faſcines over them, and preſs them down, ſo that the ſtakes of the gabions may keep them firm; then another is laid over the former, and faſten'd to them with pickets.

When the firſt ſapper is tired, he retires behind, and the ſecond takes his place; they continue thus alternately to relieve each other, that they may e-qually ſhare the danger and labour.

As this work is carried on with all poſſible ſpeed and diligence, in two or three hours, the ſappers will be tired; then the other four of their Brigade, who furniſhed them with materials, take their

places,

places, and continue the ſap in the ſame order and manner as the firſt; who furniſh the others in their turn with the neceſſary materials: Theſe four being tired, the others take their place, and ſo on alternately, till the Brigade is reliev'd, which ought to be done every ſix hours.

Not only both Approaches at the right and left are carried on at the ſame time, but alſo the ſecond Parallel to the right and left of theſe Approaches; and when thoſe who work in the Parallel come to the capital produced of the ravelin, the Approaches to the ravelin are alſo carried on at the ſame time, ſo that the ſappers work in five or ſix places at once; and ſpecial care ſhould be taken that they do not want any thing neceſſary, that the work may be continued with all poſſible ſpeed, and without interruption.

When this work is carried on to ſome diſtance, other workmen are ordered to widen the ſap to 12 feet, and from thence forward it changes its name of ſap to that of Trenches.

As this kind of work is dangerous, and requires skill in the workmen, they are paid according to meaſurement, more or leſs, in proportion as it is dangerous, from half a crown a fathom in length, to two guineas; and if there are any kill'd, the ſurvivors of the Brigade receive their pay; by which means it has ſometimes happened that one or two men have received the pay of 24 at the rate of a guinea a-piece. And tho' this kind of work may ſeem to go on but ſlowly in compariſon to that which is carried on uncovered, yet by its continuance both night and day without intermiſſion, it is quick enough; beſides, as it is done with much

leſs

leſs danger than the other, it is generally made uſe of as ſoon as the Approaches come to the ſecond Parallel.

Of PARALLELS.

PARALLELS are uſeful on account of their joining the two Attacs together, ſo as to afford mutual ſuccour to each other; they ſerve as places of Arms for the guards of the Trenches, who by that means are under cover, and at hand to ſupport the workmen upon all occaſions. The firſt Parallel is always made within 300 fathoms of the covert-way, whether the Trenches are opened nearer or farther from the place; if it can be made nearer, it would be ſo much the better. The ſecond 140 fathoms from the firſt; or if the garriſon is weak, it is often made cloſe to the glacis. After the ſecond Parallel, the Approaches being ſo near that the Beſieged may ſally out whenever they pleaſe, without any danger of their being cut off in their retreat, it is neceſſary to make at every firſt or ſecond turn of the Approaches, a kind of half Parallels, or places of Arms; ſo that a part of the guard being placed there, may be near at hand to ſupport the workmen in caſe of a ſally or an alarm.

The third Parallel is always made near, or upon the glacis, but when the Garriſon is very numerous, and have an active Governor, or when the *Ricochet* Batteries cannot be uſed on account of ſome river, moraſſes, or the Town lying on a hill, or riſing ground, much above the level of the country; it is made within thirteen or fourteen fathom of the ſaliant angles of the covert-way; and as the ſallies of the Beſieged become more frequent than ever, this Parallel is on that account made wider and deeper,

deeper, as alſo with more care than the reſt of the works, eſpecially as the troops deſigned for the Attacs of the covert-way, and other out-works, are to be placed there; for which reaſon, its parapet is made with ſteps, that the troops may march out in order of battle without confuſion.

Sometimes it is neceſſary to make a fourth Parallel, within ſix or eight fathoms of the covert-way, eſpecially when the Garriſon is very numerous, or when all their Guns cannot be diſmounted, or when the Beſieged erect new Batteries by night, in the room of thoſe that were deſtroyed by day.

Of SALLIES.

IT has been ſhewn how the Approaches, and the other works are to be made, as far as the third Parallel, without mentioning any other obſtruction from the Beſieged, than the fire of their Guns and ſmall Arms; but it is to be preſumed, that they will not be ſo remiſs as to ſuffer themſelves to be thus ſtreighten'd without making uſe of all the means in their power to prevent it, and to prolong the Siege. And as Sallies are one of the principal, they will not fail to attempt them, and endeavour to deſtroy the Trenches as often as they poſſibly can. It will not be improper therefore, to ſhew how to prevent the effects of them, and even to render them, if poſſible, advantagious to the Beſiegers.

As the garriſon is but ſmall in compariſon to the Army of the Beſiegers, the Trenches are, or ought to be ſo well guarded, as to be able not only to reſiſt any Sally, even though compoſed of the whole garriſon, but alſo to repulſe them with the utmoſt vigour, and a conſiderable loſs; a Sally therefore never ſucceeds but by ſurprize; ſo that if the Beſiegers

ſiegers are upon their guard, eſpecially when the head of the work is at ſome diſtance from the place, it will hardly be poſſible for the Beſieged to undertake any thing againſt the Trenches. But ſhould they be ſo inconſiderate as to Sally out, notwithſtanding that the Beſiegers are appriſed of it, it will undoubtedly turn to their great diſadvantage and confuſion. The intent of Sallies is to fall upon and deſtroy a part of the Approaches not quite finiſhed, or not well ſupported, to nail up the Guns upon the Batteries, or to ſurprize a part of the guard in the Trenches; for which reaſon ſpecial care muſt be taken to watch and look out ſo well, that the enemy may not approach unperceived; and as ſoon as they are heard or ſeen, the workmen at the head of the Trenches are ordered to retire, and notice is given to the guard both in the Parallel, and in the Approaches, to be ready to receive them with a ſmart fire, which being well executed, will not fail of putting them in confuſion, and then they muſt be immediately charged both in front and flank, taking care however not to purſue them too far, for fear of the fire from the covert-way, and other out-works of the Place, as ſoon as the enemies are under cover; the garriſon generally waiting for ſuch an opportunity.

Notwithſtanding all the care that can be taken, the private men will be remiſs in their duty; and therefore, an officer, ſerjeant, or corporal, ſhould continually watch over the centries, who are poſted to prevent a ſurprize; eſpecially in a dark or rainy night, or early in the morning when the ſoldiers are fatigued, and ſo leſs upon their guard; it being chiefly at thoſe times that the enemy undertakes thoſe Sallies.

The

The works become more expoſed to be inſulted in proportion as they approach the Town; for which reaſon no work ſhould be undertaken without being well ſupported, and for that purpoſe, as already ſaid, the places of Arms are made; from thence the works are maintained till ſuch time that the third Parallel is quite finiſhed, which will ſo ſtreighten the Beſieged, as not to be any longer in a condition to Sally without the utmoſt danger.

Sallies are never made in the day-time, but by a preſumptuous enemy; for then they are eaſily repulſed, unleſs the garriſon is very numerous, or the Army of the Beſiegers ſo weak, as not to be able to furniſh a ſufficient guard for the Trenches; in ſuch caſes a General ought to conſider whether he is in a condition to continue the Siege.

A garriſon may be in a ſtate to inſult or attac the Trenches after having received a ſtrong reinforcement, or when the Beſiegers are obliged to ſend a conſiderable part of their Army upon ſome other enterprize. It is then at the option of the General, either to continue or raiſe the Siege; if it appears that the Army ſuffers greatly without any certainty of taking the Place, he ought to retire.

But ſuppoſe that meaſures have been ſo well concerted by the General, as to prevent in all appearance the danger of being diſappointed; then if parties of 10 or 12 men headed by a ſerjeant, are ordered to range in the night between the Trenches and the Town, to watch, and to diſcover the enemy, in caſe they come out, and give notice to the guard of the Trenches; this will defeat their deſign. Theſe men muſt be on their faces as near the covert-way as poſſible, remaining in profound ſilence till they hear or perceive ſome motion, then

ſend

ſend one of them immediately to the guard, whilſt the reſt continue there as long as they can be concealed, to ſee which way the enemy direct their courſe, and then retire. This may be eaſily executed without much danger, and will ſecure the Beſiegers from all ſurpriſe.

When the works are advanced to the third Parallel, and before that Parallel is finiſhed, if the enemy ſhould then Sally out of a ſudden, and fall upon the workmen, they muſt be ordered to retire, to let the guard fire briskly upon them, without minding the overturning a dozen or two of gabions; for the galling fire of the ſmall Arms, will ſoon oblige them to retire, and then the workmen muſt return inſtantly and repair their works, which may ſoon be done; whereas the Beſieged are not in a condition to repair their loſs ſo quickly.

Of the Lodgment *on the* Glacis, *and taking of the* Covert-way.

Pl. X. IT has been ſhewn how the Approaches are conducted as far as the third Parallel; but it is hardly poſſible to proceed any farther, without being ſeen or enfiladed by ſome work or other of the Town; it is neceſſary to make the Trenches from thence forward, much deeper than before, with traverſes in them, from diſtance to diſtance; by which the enfilades will be prevented in part, tho' not entirely.

The figure of theſe kind of Trenches varies according to the different circumſtances or poſition of the works which defend the glacis: Sometimes they are made partly with ſhort turns, or Zig-zags, like the former, and the reſt with a direct ſap over the ridge

ridge of the glacis, before the faliant angles of the covert-way, or otherwife, by making a large oval traverfe, as in this Plan; and then they are carried directly forward over the ridge of the glacis as aforefaid, in the following manner.

Two fappers roll each mantlet, or ftuft gabion before them, each carrying on a fap, one on one fide of the ridge, and the other on the other, fo that the interval between the Parapets may be ten or twelve feet; and then they make traverfes from diftance to diftance of about eight or ten feet thick, with a paffage round each end, covered by two or three gabions.

This work is called Double Sap, the Plan of which fee in Plate V. When it is advanced within thirteen or fourteen fathoms of the faliant angles of the covert-way, that is, within throw of grenades, two branches are made to the right and to the left, fuch as a, which M. *Vauban* calls *Cavalliers* of the Trenches: They are nearly perpendicular to the produced branches of the covert-way fo as to enfilade them, of about twenty fathoms long, eighteen feet wide, with their Parapets raifed by three ranges of gabions placed in fteps-fafhion, fo that from thence the covert-way may be plunged into every where; they have little turns at the end, to prevent being enfiladed by the places of Arms in the re-entring angles.

M. *Vauban* judicioufly obferves, that the Approaches carried along the ridge before the faliant angles are only enfiladed by three or four men placed in thofe angles, and therefore this is the fecureft place to make the Approaches. From thefe places of arms or *Cavalliers,* the double fap is continued quite up to the angles of the covert-way, in the

the ſame manner as before; and to prevent the enemy from throwing ſtones and grenades into them, they muſt be covered with blinds and faſcines.

When theſe *Cavalliers* are once finiſhed, it will be dangerous for the enemy to remain any where in the covert-way, except behind the traverſes, as they will be every where elſe ſeen and enfiladed from them. I ſuppoſe that the *Ricochet* Batteries are playing continually into the covert-way, and adjacent works, whilſt the lodgments are carried on upon the glacis, otherwiſe the enemy will obſtruct them to ſuch a degree, and fire upon them ſo briskly, that it would be almoſt impoſſible to advance a ſtep. Here a gallant Governor may with the greateſt advantage oppoſe the Beſiegers, and diſpute every inch of ground to the laſt, as being ſo near the covert-way, where his troops may do them great miſchief with little or no danger, were it not for the *Ricochet* Batteries, and conſequently this is the place where the Beſiegers may make the beſt uſe of thoſe Batteries.

The Approaches being advanced to the ſaliant angles of the covert-way, a large oval traverſe r is made, ſo that the trench round it may take in a part of that angle, by which the enemy may be drove out of that place; from whence the Trenches are continued both ways along, and within three fathoms of the covert-way towards the places of Arms.

The ſaliant angles of the covert-way being taken, and lodgments made in them, the Beſieged muſt be drove from behind the traverſes next to them; by piercing into the paſſages behind them, from the

the oppofite Trenches, and making lodgments there, which are afterwards carried round the angle of the counter-fcarp. This done, there will be no place left to the Befieged in the covert-way befides the places of Arms at the re-entring angles.

When the Trenches are advanced to about two thirds of the way, branches are carried on both fides of the places of Arms, fo as to meet, and that both together make a right line nearly, in which Batteries, fuch as b, for throwing of ftones are made; which, when finifhed, will ferve to drive the Befieged out of the oppofite places of Arms. But when there are ftone redoubts in thefe places, which are covered, the ftones will have no effect on them, and therefore fhells muft be thrown to ruin them, which when done, a detachment of about 200 grenadiers is fent to attac the Places of Arms fword in hand. So foon as the Befieged are retired, workmen headed by Engineers are fent to make lodgments in them, which are afterwards joined to thofe on the glacis, as likewife to thofe which are carried round the places of Arms.

If the garrifon is ftrong, and the Governor has courage and conduct, he will not let the Befiegers make thefe lodgments quietly, but on the contrary, will return as often as he can, to difturb and deftroy them, till they are quite finifhed, and a fufficient number of men placed to defend them.

Thefe places of Arms being once taken, and the lodgments in them finifhed, the Befieged will hardly venture to return any more; Batteries are made along the gorges, in a round form, to fire upon the tenailles, curtain, and the faces of the oppofite Baftions.

Of

Of Mines.

THE Art of Mining being independant of the Attac and Defence of Places, I ſhall treat of it more diſtinctly afterwards, and only mention here a few particulars, ſo far as relates to the Attac, without entering into the manner of loading and making the Mines.

What has been ſaid relating to the works made on the glacis and covert-way, is only with regard to thoſe above ground, and to the reſiſtance the Beſiegers meet from the enemy in thoſe works. But as there are few Places which have no Mines, and when the Beſieged are obliged to yield to the ſuperiority of the Beſiegers, both in men and guns, and all hopes of recovering the loſt works, they will not fail to ſpring their Mines, as the laſt and only reſource left to retard the progreſs of the enemy; and thereby blow up both them and their works; by which they oblige the Beſiegers to attac thoſe works again, and make new lodgments therein, which they thought ſecure before. It is therefore neceſſary to ſhew how to find out their Mines, to prevent their effect, or deſtroy the greateſt part of them, and ſo be upon a par with the Beſieged in that reſpect.

To effect this, ſhafts or pits are ſunk in or near the third Parallel, of 7 or 8 feet ſquare, and about 18 or 20 deep, if the ground will permit it; from thence galleries are carried on towards the covert-way, of 4 feet wide, and 5 high, endeavouring to meet with thoſe of the enemy, by boring the ſides, bottom, and roofs of theſe galleries with a long iron needle or augre, from time to time, to find

find out whether the enemy's galleries or miners are near; if they are found to be underneath, an opening is made down into them, and live ſhells thrown in, either to drive them out, or deſtroy them; if on the contrary they are found to be above, a ſmall mine muſt be made to break them; but if they are at one or the other ſide, notice muſt be taken whether they approach or recede. In the firſt caſe, a hole is made, and a piſtol clapt in it, ready to fire at them, ſo ſoon as they are within reach; and in the ſecond caſe, a gallery muſt be carried on towards them, within a ſmall diſtance, and then a ſmall Mine ſhould be made, to burſt their gallery, and deſtroy the miners.

The Beſiegers ſhould carry on their galleries directly under their works above ground, to ſecure them from the enemies Mines; and if thoſe of the Beſieged are not diſcovered, as it may eaſily happen, branches are made on the right and left, with ſmall Mines at the ends of them, which being ſprung, will hardly fail to deſtroy moſt of the Beſieged's Mines and Galleries.

Notwithſtanding all the care that may be taken, it is not to be preſumed, that all the Mines of the Beſieged ſhould be rendered ineffectual, and therefore, ſo ſoon as they ſpring any of them, workmen are immediately ſent to lodge themſelves in the pits. If water could by any means be conveyed in the Beſieged's galleries, it would effectually ſpoil them, as it happened at the Siege of *Turin*, where many Mines were rendered uſeleſs that way.

The Beſieged generally place Mines within 4 or 5 fathoms of the covert-way where they know that the Beſiegers will lodge themſelves, not daring to make them nearer, for fear of breaking the Palliſades;

ſades; the Beſiegers ſhould therefore make Mines thereabouts alſo, to render the others uſeleſs, and thereby make partly their own lodgments. This precaution ſhould in my opinion be taken wherever there is a ſuſpicion of any Mines of the Beſieged, which conſequently would prevent any great miſchief from them, eſpecially as their effect is but inconſiderable, except directly upwards only.

The Beſieged will endeavour alſo to the utmoſt of their power, not to part with an inch of ground till conſtrained by the laſt neceſſity; and when they have done all they can above ground, to maintain it, will at laſt ſpring their Mines; for which reaſon the Beſiegers ſhould act with the utmoſt caution, not to be ſurprized, and to get maſters of as much ground with as little loſs as poſſible.

If notwithſtanding all that has been ſaid about taking the covert-way, the Beſieged ſtill keep their ground, in the places of Arms, and behind the traverſes, the Beſiegers muſt endeavour to oblige them to ſpring their Mines near the ſaliant angles, or elſe muſt make Mines there themſelves, and blow them up; and ſo ſoon as this is done, ſend workmen immediately to make lodgments in their ruins, ſecuring thoſe, as much as poſſible, which are joined afterwards to the double ſaps on the ridge before the ſaliant angle, and particular care muſt be taken to make traverſes every where, to ſecure them from being enfiladed. When theſe lodgments are extended to the traverſes next to the ſaliant angles, and the enemy ſtill keep their ground behind them, as there can be but few under cover there, a party of grenadiers muſt ſet upon them on a ſudden to drive them away, ſword in hand; this done, ſome of theſe grenadiers ſhould endeavour to find out the

the entrance of the Beſieged's galleries, and cut off the matches, to render them uſeleſs; in the mean time workmen are ſent to lodge themſelves in the paſſage round the traverſe, and from thence carry a lodgment round the counterſcarp before the ſaliant angles, till they meet each other; theſe lodgments muſt be made very deep, and cover'd with blinds and faſcines, to prevent the effect of the grenades and ſtones that may be thrown in them by the Beſieged.

Whilſt this is doing, the cannons and mortars muſt inceſſantly fire upon all the defences of the Place, endeavouring to ſilence the fire of the Beſieged, and to drive them, if poſſible, from behind the parapets of the faces of the oppoſite baſtions and ravelins.

ATTAC *of the* COVERT-WAY, *ſword in hand.*

WHEN a Town is not ſtrongly garriſoned, or the General thinks the foregoing method too tedious, the Attac may be made as follows.

The third Parallel in this caſe ſhould be made at leaſt as forward as the midway of the glacis, having its parapet made ſtep-faſhion, as before, that the troops deſigned for the Attac may paſs eaſily over it, without any confuſion. A great quantity of faſcines, gabions, and other materials, muſt be got ready, and placed at the back of this Parallel. A ſtrong party of grenadiers is ordered, and placed in this Parallel, five or ſix deep, and the workmen behind them on the reverſe of the Parallel, having their tools and materials by them. Likewiſe all the adjacent parts of the Trenches muſt be well furniſhed with troops to ſupport the grenadiers, if there

there is occaſion, and fire wherever the enemy appears. The grenadiers muſt be provided with hatchets to cut the palliſades, in caſe the guns ſhould not have broke them.

Before the Attac is made, the guns and mortars are to fire briskly for ſome time at all the defences of and into the Covert-way, to drive the Beſieged from thence, and to break the palliſades, if poſſible, and to plough the ridge of the glacis in ſuch a manner, as the troops may enter the Covert-way without much difficulty; then the guns ceaſe, to cool. When this is done, the ſignal is given for the Attac, upon which all the troops begin to move, and paſſing quickly over the parapet of the Parallel, march directly to the Covert-way, which they enter either thro' the ſally-ports or paſſages made by the guns, or elſe the grenadiers cut down the paliſades with their hatchets; and being entered charge the enemy ſo vigorouſly as to oblige them to retire; then the Engineers ſet the workmen about making a lodgment on the ridge of the glacis, oppoſite to that part of the Covert-way which the Beſieged have abandoned.

Theſe lodgments are made with gabions and faſcines, in the ſame manner as the ſaps; and traverſes are made every where to prevent the enfilades. The troops keep behind the workmen, and kneel down, till the lodgment is ſo far advanced that they may retire into it. Whilſt this is doing, the Batteries fire continually upon all the defences of the Covert-way, either to ſilence or abate the fire of the enemy, as much as poſſible, and to oblige them to think more of their own ſafety, than oppoſing the Beſiegers.

If

If the Beſieged ſhould return the charge, as probably they will, and overthrow the work and maintain their ground, nothing but a ſuperior force can make the Beſiegers maſters of the Place.

When the Beſieged find that they cannot poſſibly hold out any longer, they will ſet fire to their Mines and retire; upon which workmen are immediately ſent to make a lodgment in their ruins, which is afterwards joined to the reſt of the Trenches.

This was the manner of attacking the Covert-way formerly; but ſince M. *Vauban*, by his great experience and knowledge, has brought the Art of Attac and Defence to ſo great a perfection as it now is, the Covert-way has ſeldom been taken ſword in hand. His chief ſtudy being always to preſerve the troops as much as poſſible, and never to expoſe them to any danger, without the utmoſt neceſſity.

Yet, when a garriſon is but weak, and the Army of Beſiegers very ſtrong, the guns of the Beſieged may be ſilenced, and the paliſadoes torn to pieces by the *Batteries Ricochet;* in ſuch caſes the Covert-way may be attacked with open force, ſword in hand, and that without much danger; but if the garriſon is ſtrong and commanded by a Governor who knows his buſineſs, it would be imprudent to make ſuch an Attac; for it would prove one of the moſt bloody actions of the whole Siege.

The *French* loſt 3000 men in the Attac of the Covert-way, ſword in hand, in the laſt Siege of *Fribourg* in *Briſgaw*.

BAT-

BATTERIES *on the* COVERT-WAY.

THE Befieged being drove out of the Covert-way, the next thing to be done is erecting Batteries, to make a Breach, and ruin the defences.

Pl. X. As the ravelin C is to be attacked firft, and is flanked or defended by the faces of the baftions A, B, oppofite to its ditch; the firft Batteries to be made are thofe marked E, F, the former to ruin the defence of the ditch, and the latter to make a breach in the ravelin. Each of thefe Batteries fhould contain four or five twenty-four pounders at leaft: fometimes the Batteries, F, F, which batter the ravelin, are alfo ufed to batter the faces of the baftions A, B; but as this cannot be done without battering the ravelin too near the faliant angle, it will be much better to erect Batteries for that purpofe.

If the faces of the baftions, which defend the ditch of the ravelin, were well enfiladed by the *Ricochet* Batteries, there will be no occafion for the Batteries, E, E.

Whilft thefe Batteries are erecting, others alfo, as G, H, muft be made to ruin the flanks of the baftions A, B, of as many guns as can be placed oppofite to them, as alfo fome to batter the oppofite faces.

It is to obferved, that the beft place for making breaches, both in the ravelin and baftion, is about 15 fathoms from the faliant angle; fo that battering on both fides of the angle, the breach will become fpacious and large; whereas, on the contrary, if the breach were made too near the angle, it would become too fmall and incommodious.

The

The Batteries made in the gorges K of the places of Arms, ſerve to fire obliquely at the faces, curtains, and tenailles, as mentioned already, or to place mortars, for throwing ſhells and ſtones into the works.

When the Batteries are erected, paſſages over the trenches are made to bring the guns there; which when done, are again opened, ſo as not to interrupt the free paſſage of the Trenches.

All theſe Batteries are made on the ridge of the glacis, within three fathoms of the covert-way, which interval ſerves to make the parapet or epaulement; and as the guns, which are to make a breach, ſhould fire as low as the bottom of the ditch, the embraſures muſt be made ſloping downwards accordingly. The guns intended to make a breach ſhould always fire together, or in ſalvos, and as nearly in the ſame place as poſſible, and continue ſo till the wall is quite well pierced, which may be known by the earth's falling. They muſt begin firing at firſt as low as they can, and batter the wall nearly in a horizontal line, as far as the Breach is intended to be made; afterwards fire ſomething higher, in the ſame manner as before, and continue ſo to do, till the wall tumbles down; for ſhould they fire at firſt too high, the rubbiſh would cover the lower part of the wall, ſo as not to be broke afterwards, and the Breach would thereby become impracticable.

It was formerly the cuſtom, to load the battering Pieces with a quantity of powder, equal to half the weight of the ſhot; but experience has ſhewn that one third of the ſhots weight, and perhaps leſs is ſufficient. For the *French* uſed no more in the laſt war; and it is manifeſt that a ſhot with a very great velocity,

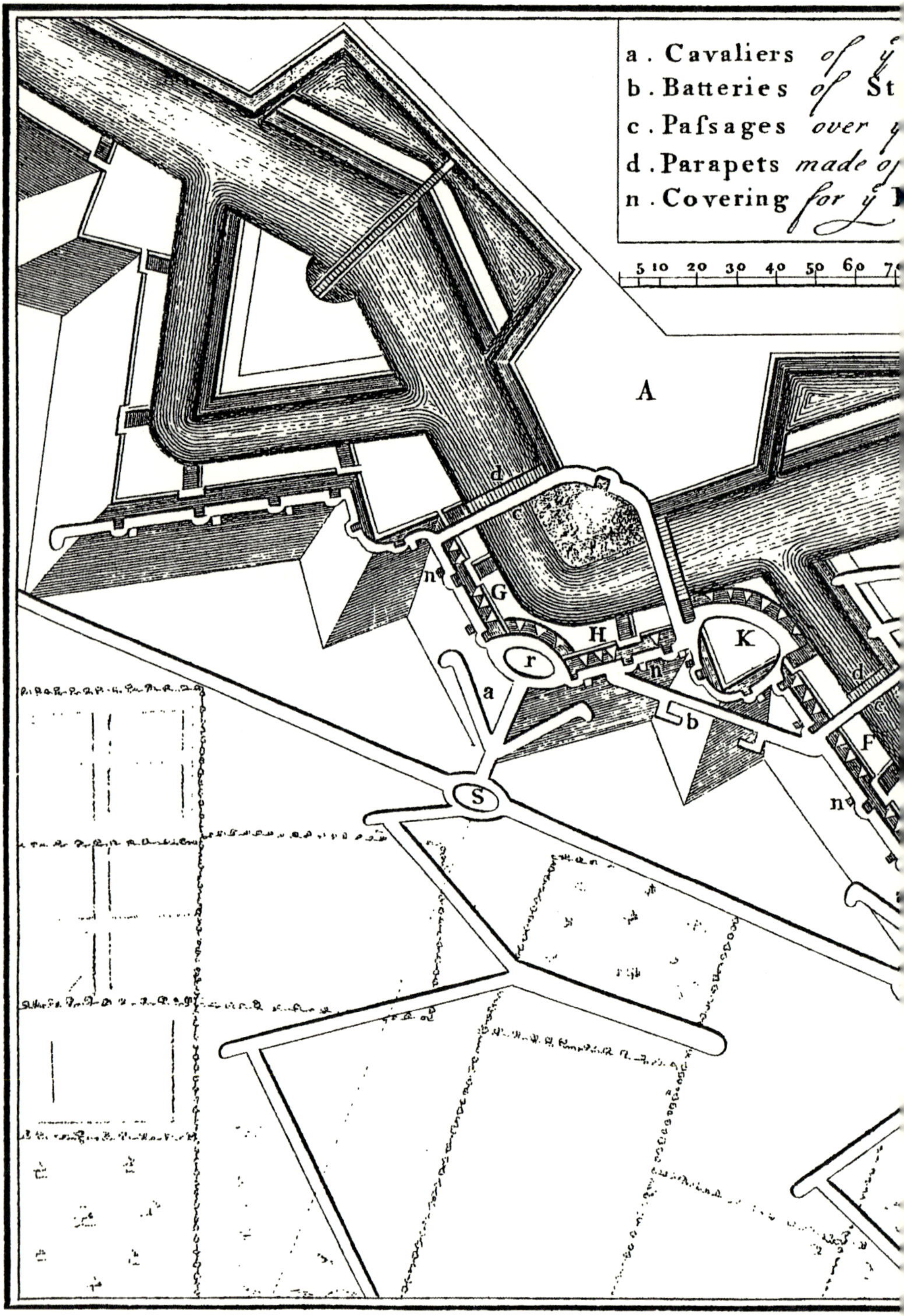

a. Cavaliers of y
b. Batteries of St
c. Pafsages over y
d. Parapets made o
n. Covering for y
5 10 20 30 40 50 60 7
A
d
c
n
G
H
r
K
n
a
b
d
c
F
S
n

X.

renches .
& Shell Mortars .
itches .
ascines to Cover ye Passage
ages round ye Traverse
120
160 Fathom
B
G
K
H
n
a
b
S
d
c
F
n

velocity, enters the wall without ſhaking it, whereas if the velocity is juſt ſufficient to enter the wall, it makes a larger hole, and ſhakes it very much at the ſame time. This is confirmed by experience; for if a ſtone is thrown againſt a window or any other glaſs, it will break it into many pieces, but a piſtol ſhot will only make a ſmall hole without breaking it into pieces.

It has likewiſe been found, that battering-pieces need not be ſo heavy as they have been made hitherto; for inſtead of making a twenty-four pounder to weigh 5100, if it weighs 2400 it is ſufficient to reſiſt all the firing in a Siege, which diminiſhes both the charges, and the trouble of bringing them to the Battery, conſiderably, as ſhewn in my *Treatiſe of Artillery.*

The modern and ſhorteſt way to make a Breach, is to pierce the wall with guns, ſo as to make a hole for two or three miners to lodge themſelves in it, who are ſent over the ditch, as ſoon as it is dark, in boats, if the ditch is full of water, and make three or four ſmall Mines under the parapet, which when ſprung, will at once make a Breach ſufficiently large, and wanting nothing but to be cleared, in order to make an Aſſault.

As the Beſieged will not fail to endeavour to deſtroy the Batteries near the covert-way, where they have generally mines ready made, it would be prudent to make galleries quite round them, either to deſtroy thoſe mines, or to render them uſeleſs.

DESCENTS *and* PASSAGES *over* DITCHES.

WHilſt the Batteries on the glacis are erecting, preparations are made for the paſſage over the ditch of the ravelin; which may be dry or wet;

wet; and either of them requires a different manner of paſſing. I ſhall explain both ways ſeparately in their order.

If the ditch is dry and pretty deep, the deſcent is made by galleries, or a covered ſap, underground, having its entrance about the middle of the glacis, and coming out near the bottom of the ditch, oppoſite to the Breach; theſe galleries are made with wooden frames, and boards to ſupport the earth at the ſides and roof, and to prevent its falling in. PL. XI.

To find the ſlope of the deſcent, ſo as to come out nearly at the bottom of the ditch, its depth muſt be taken near the counterſcarp with a plummet, and the diſtance from the entrance of the gallery to the ditch muſt likewiſe be meaſured; then the horizontal diſtance of any part from the entrance muſt be to the vertical deſcent, as the length of the diſtance from the ſaid entrance to the ditch is to its depth. For inſtance, if the depth of the ditch is 30 feet, and the diſtance from the entrance of the gallery to the ditch 120, then by advancing four yards, the ſlope muſt ſink one; as 120 is to 30, ſo is four to one. But if the ditch be only 12 or 15 feet deep, then, inſtead of a gallery made underground, the deſcent is made by a ſap only, beginning in the covert-way, ſunk in proportion as it advances, ſo as to come out nearly at the bottom of the ditch. The ſides of this ſap are alſo ſecured from falling in by boards and ſtakes, and covered with blinds and faſcines.

The Beſieged will endeavour to diſturb this work, by frequent ſallies, and by throwing ſtones, ſhells, grenades, and whatever elſe may ſerve their purpoſe;

purpofe; efpecially at the entrance into the ditch. The Befiegers muft therefore take fuch precautions as may render thofe endeavours ineffectual, which may be done by throwing fhells and ftones from their Batteries, and grenades from the adjacent works into the ditch, and wherever the enemy appears; and alfo to fire at them with cannon and fmall arms; and as they are fuperior both in refpect to men and artillery, they will, in fpite of all obftruction, be able to continue their work.

When the Defcent is made, the paffage is continued with fapping, and the fide next to the place is covered with a ftrong parapet.

There are generally two or three defcents made into the fame ditch, at once, to help and fupport each other; and as the Befieged frequently make ufe of fmall mines to blow up the faps, and fally out at the fame time, to deftroy and retard the work, fome companies of grenadiers are placed near at hand, to fall upon them as foon as they appear, the Batteries playing continually at all the defences of the place, that thofe within may have enough to do to think of their own fafety, without being in a condition to annoy the works of the Befiegers.

Paffages are made on both fides of the ravelin, as may be feen in Plate X. fo that the Befieged being affaulted on both fides at once, with a great number of men, and thereby overpowered, they will be obliged to retire, and leave the Befiegers mafters of the Breach.

If the ditch is wet, and within three or four feet full of water, the defcent muft begin near the Counterfcarp, fo as to come out pretty near even with the furface of the water. It is made with a covered

covered ſap in the ſame manner before deſcribed, and a great quantity of faſcines are thrown into the ditch, with ſtones or ſome other heavy materials faſtened to them, to make them ſink to the bottom, and to prevent their being carried away by the current, if there is any. When as many are thrown in, that they come above water, the ſappers range them into beds, with their forks and hooks, as well as they can; then they throw earth upon them, and make them ſmooth and fit to walk conveniently over. The beſt means to haſten a work is, to place a good number of men, at four or five feet diſtance from one another, to forward the faſcines and other neceſſary materials from hand to hand, which will prevent all that confuſion, that would otherwiſe be unavoidable, in the paſſing and repaſſing of ſo many men. PL. XII.

A ſtrong parapet of faſcines and ſand-bags filled with earth is raiſed on the ſide next the place, and if the paſſage ſhould happen to be over-looked or plunged into by any part of the place, the foremoſt workmen muſt be covered with a great heap of faſcines, which are puſhed forwards as the work advances; and as faſt as the paſſage goes on, it muſt be covered overhead with blinds and faſcines, having earth thrown over them only as far as neceſſary.

But if the ditch is full of running water, the paſſage will be more difficult; for beſides the precautions mentioned before, the faſcines muſt be loaded with beds of ſtones, heavy pieces of wood, and other materials, to prevent their being carried away by the current, and as ſoon as the work appears above water, the faſcines muſt be fixed, by driving long ſtakes thro' them into the bottom.

Note,

Note, If the current of the water is ſo violent as when ſtopped to occaſion ſuch ſwellings as might overturn the works in the ditch, then when the paſſage is advanced within 15 or 20 feet of the Breach, pieces of timber are laid a-croſs, and ſtrong planks over them, by which means a paſſage is preſerved for the current, and the works are ſecured.

ATTAC *of the* RAVELIN.

WHILE the deſcent and paſſage over the ditch are going forward, the Breach is making likewiſe, ſo that they are both finiſhed about the ſame time. If the Breach is made with cannon, care muſt be taken to begin to break the wall at the bottom of the ditch, as was obſerved before in ſpeaking of the Batteries. The wall being broke down, as far as is neceſſary, the counterforts muſt be beat down likewiſe, which as they advance farther into the earth than the wall itſelf, it will be ſooneſt done by firing at them obliquely. But if the Breach is made by Mines, which is the ſhorteſt way, the wall muſt be firſt pierced with cannon, and then miners ſent to enter the hole, as above.

The Breach being made, and the ground levelled, materials and tools at hand, the troops for the Aſſault are ordered to be in readineſs at the firſt notice, and others ready to ſupport them in caſe of need; the guns and mortars are ordered to fire briskly on all the defences of the Place, which flank the front attacked, both directly and enfilading them by the *Ricochet* Batteries; and a great number of ſhells and ſtones are thrown into the ravelin and its defences. Thus having continued for

for ſome hours, the ſignal is given for ceaſing to fire; then the grenadiers march to the Aſſault, charge the enemy vigorouſly, and oblige them to retire; after this the workmen, headed by an engineer, advance to make a lodgment on the upper part of the Breach, which when partly done, the troops retire behind it, ready to repulſe the Beſieged if they ſhould return, which it is preſumed they will not fail to do ſo long and often as they can: but when this lodgment is once ſo far advanced, as to be in a condition of receiving a good number of troops, the Beſieged will have it no longer in their power to diſturb them by open force, and therefore they will make uſe of their Mines, as the laſt reſource left them, if they have not been prevented before-hand; either to blow up the lodgment, or any other place where they perceive a body of men.

When the ditch is dry, the Beſiegers may prevent the effect of the Beſieged's Mines, by ſending ſome ſtout grenadiers, with promiſe of paying them well, from the lodgment on the counterſcarp, thro' the ditch round the ravelin to its gorge, at the ſame time the Aſſault is making, to diſcover their entrances, and to cut off the match, whereby the Mines will become uſeleſs.

Another way, and more in uſe, to get poſſeſſion of the Breach is, by ſending two or three ſappers to the extremity of the Breach next to the place, where is generally room behind the break of the wall to cover one or two men; there they begin a lodgment for themſelves and for two or three more, which are ſent after them, who, together, make room for others to follow; and in this manner they proceed, ſtill enlarging their lodgment till the whole

is

is compleat. It is to be prefumed that the Befieged will oppofe this work with all their power, either by falling fuddenly upon the workmen, or throwing great quantities of grenades, and all kinds of fire-works amongft them. To prevent all this, the workmen are ordered to retire upon the firft appearance of an enemy; and the Batteries, as likewife the troops in the lodgments at hand, fire at the fame time all together upon them, which will foon oblige them to retire; upon which the workmen return immediately to their work again; and this way of proceeding muft be continued till the lodgment is quite finifhed. If the enemies fpring any Mines, workmen are directly fent to lodge themfelves in the cavities made by them; and from thence they extend themfelves to the former lodgment, or to that part of it which has not been overthrown.

The lodgment at the upper part of the Breach being completed, others are carried on by faps along the parapet, to get poffeffion of the Pl. X. traverfes, if the Befieged have made any to fhelter themfelves from the ricochets, as alfo to enfilade obliquely the paffage between the tenaille and the ravelin. Thefe lodgments muft be well traverfed from the enfilades to which they are unavoidably expofed, and joined by others crofs the ravelin, which ferve as a fort of parallels, as likewife to diflodge the enemy from the retrenchments they may have made in the gorge of the ravelin, and to enfilade the communication between the ravelin and the body of the place.

If there fhould be any palifades placed either round the befieged's retrenchments or ferving for retrenchments themfelves, it will be neceffary to break

break them with the guns from a Battery that can ſee them; or if that cannot be done, ſmall Mines are made to blow them up; and then the ſaps are carried on to the gorge of the ravelin, to prevent the Beſieged from returning into the work.

The ATTAC *of the* BASTION.

WHILST the Attac of the ravelin is making, that of the Baſtion is carried on at the ſame time; the deſcent and paſſage over the ditch and the Breach, are made much in the ſame manner as thoſe of the ravelin; but as the ditch is here much wider and deeper, as likewiſe better defended by the flanks, than there; the making the deſcent and paſſage over the ditch, is alſo attended with much more difficulty and danger: beſides, this being the laſt place the Beſieged have to defend, and where they have all their ſtrength united, they will exert themſelves to the utmoſt of their power, to obſtruct and retard it as long as poſſible: it will not therefore be improper to add ſomething to what has been ſaid already on that account, eſpecially when the ditch is full of water, and has a ſtrong current.

If the ditch is very deep, and the current either rapid of itſelf, or is to be made ſo by means of ſluices, the paſſing it is certainly the moſt difficult part of the Attac, unleſs there is a poſſibility of turning the current into ſome other channel. If the water is kept up by ſluices, they muſt be found out and deſtroyed if poſſible, either by ſhot or ſhells; in ſhort, no means muſt be left untry'd to break them; and if that can be effected, it will

ex-

extremely facilitate the paſſage. * But if the ſluice is not to be broke by any means whatſoever, there will be no other remedy than to make a ſtrong dyke croſs the ditch, ſo as to keep up the water to the ſame height as the ſluice does: this dyke is made by laying firſt a broad bed of faſcines, having ſtones tied in them ſo as to ſink to the bottom, and taking care to make it of ſufficient breadth to be able to reſiſt the current; over theſe faſcines are thrown a great quantity of earth and ſtones mixt together, and the whole beaten or rammed down, and long ſtakes driven thro' it, if neceſſary, to make it ſufficiently ſtrong.

When this work is advanced within five or ſix fathoms of the Breach, the current being then confined to a narrow paſſage, will become extremely rapid; whatever comes to mind is then made uſe of to finiſh it; as ſtuffed gabions, great ſtones, blocks of wood, old leaky casks, boats or ſmall craft, &c. if any can be had, and pieces of timber for a bridge laid over them.

All this is to be done, and the work carried on within two or three fathoms of the wall, before the Breach is made; then the guns firing againſt the lower part of the revetement, which falling in the ditch, helps to fill up what remains: but if that ſhould not be ſufficient, miners are ſent over to blow up as much more earth as is wanting.

If the ditch be deep and narrow, miners are ſet to work on both ſides of it, who by making their mines pretty deep, and loading them with a great deal

* At the Siege of *Aeth,* in the year 1697, Marſhal *Vauban* throwing large ſhells of 500 lb. found means to break the ſluice which kept up the water of the ditch in the front of the Attacs; the paſſing which would otherwiſe have been attended with much trouble, but was by this means made eaſy.

deal of powder, will fill up the greateſt part of the ditch with rubbiſh.

If there are tenailles before the curtains, the Batteries placed in the places of Arms of the covert-way, muſt endeavour to ruin them with their guns, throwing likewiſe a great number of ſhells and ſtones into them: and if the Beſieged ſhould make any embraſures in the upper curtain to fire obliquely at the paſſage over the ditch, a Battery of guns muſt be made in the ravelin near the ſaliant angle, which will command not only the whole curtain, but alſo the paſſage between the tenailles; and if this is not yet ſufficient to ſilence the fire of the place, another Battery of mortars muſt be made in the gorge of the ravelin, from whence ſuch ſhowers of ſtones and ſhells may be thrown into all the works, that the Beſieged may find but very little ſhelter there.

But if notwithſtanding all this, they maintain their guns behind the curtain, as being cover'd by means of the obliquity of the embraſures; it will be neceſſary to find a proper ſituation to erect ricochet Batteries for enfilading the curtain, which is not eaſily done without having a good Plan of the place; for it is a hard matter to know in what line the curtain lies, and no place but a perpendicular to it will do. But ſuppoſing the paſſage over the ditch and the Breach made ready to march over, the Beſiegers muſt endeavour to eſtabliſh a good lodgment in the upper part of the Breach, much after the ſame manner as has been explained in the Attac of the ravelin; that is, by ſending ſmall parties thither to lodge themſelves, with orders to retire as ſoon as the enemy appears.

If

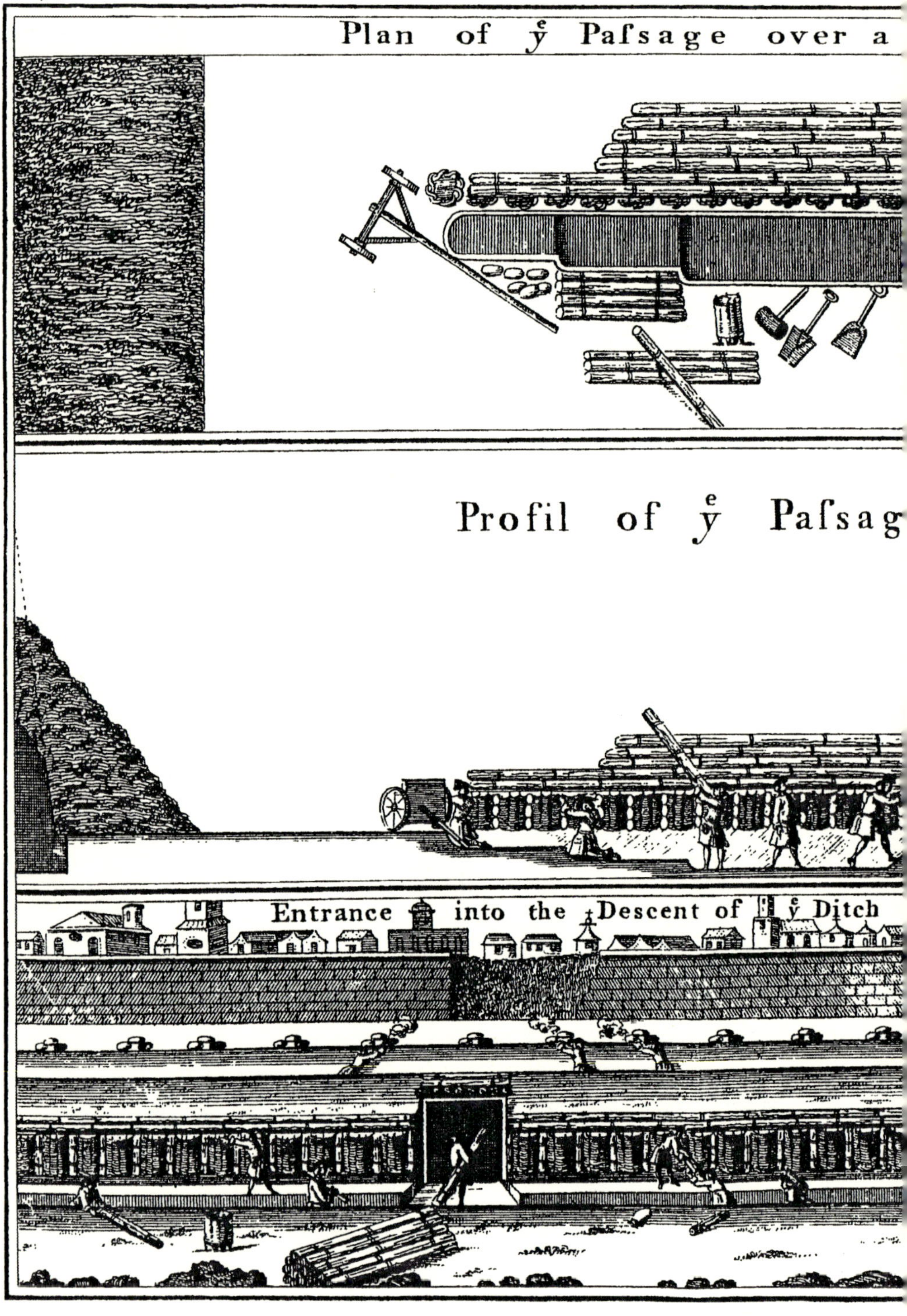

Plan of ye Pafsage over a
Profil of ye Pafsag
Entrance into the Descent of ye Ditch

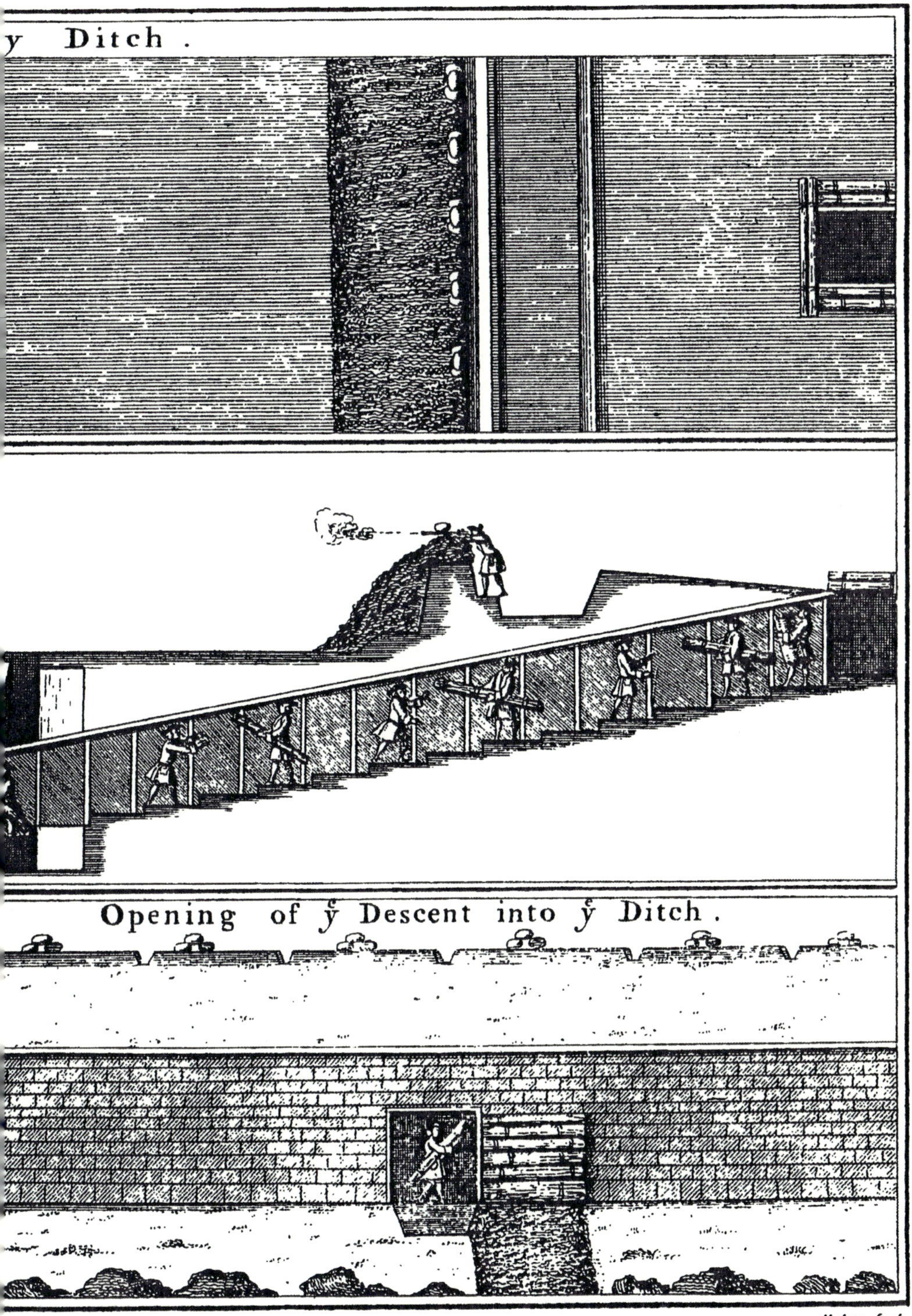

Hulett Sculp

If the baſtion has a retrenchment with revetements, the Beſieged will defend the Breach with great obſtinacy; in ſuch caſes a general Aſſault muſt be made on both ſides of the baſtion at the ſame time, with a ſufficient number of men to force the enemy to retire, and theſe muſt be ſuſtained by large detachments, poſted on the right and left near the deſcent; and as ſoon as the Beſieged give way, the Beſiegers lodge themſelves in the Breach, which being finiſhed, ſaps are carried on to the right and left along the parapet towards the gorge of the baſtion, as likewiſe others towards the flanks or orillons, if there are any, from whence the Beſieged may be entirely driven out of the tenailles, if they have not quitted them before.

If the baſtions are detached from the body of the place, it will probably be neceſſary to erect a battery of cannon in the baſtion, to batter the inſide wall, as likewiſe to place ſome mortars there, for throwing ſhells and ſtones, under the protection of which, the Approaches may be carried on quite up to the inſide ditch, and then a Breach is made in the body of the Place, either with cannon or Mines; this done, the Beſieged will hardly venture to ſtand an Aſſault, as being too dangerous for them, conſequently we may reaſonably ſuppoſe that they will beat the Chamade.

Of MINES *under the* BASTION *and* RAVELIN.

WHAT has been ſaid relating to the Attacs of the Baſtion and Ravelin, was only in regard to the works above-ground; it remains now to ſpeak of thoſe made under-ground, which are conſiderable enough to be treated ſeparately.

To

To fix the miners to the walls when the ditch is dry, a lodgment is made on the oppofite counterfcarp near the defcent, to protect from thence the miners againft the fallies of the Befieged; then the wall is broke with the cannon as near the bottom as can be done, to lodge the miners, and to get under the galleries of the Befieged if poffible: the miners remove the rubbifh to make room for lodging two or three of them in the hole. Sometimes when the ground is favourable, they carry their galleries from the covert-way under the ditch to the foot of the wall. But if the ditch is wet, the wall is pierced with cannon, while the paffage is making, and the miners get over in boats in the dusk; but as foon as the Befieged difcover this, they will endeavor to deftroy the miners, by throwing great quantities of fhells, grenades, and all kinds of fireworks before the hole; for which reafon the miners muft make all poffible hafte to get under cover; that will foon be done, if the wall has been thoroughly pierced by the cannon.

If there is any fufpicion of Counter-mines behind the wall, the miners retire fo foon as the rubbifh is removed, and the cannon fire again into the hole, endeavouring to break the galleries. If this can be done, the miners proceed with their work; but in cafe the guns have not reached them, the beft way then is, to make fome fmall Mines at the right and left of the hole, which when fprung, will not fail of deftroying the enemies galleries, and then the miners prepare their Mines for making Breach.

The miners relieve one another every two hours, to carry on their works with more fpeed, which the Befieged, on their fide, will endeavour to obftruct, by by repairing their galleries, and lying in wait for the Befiegers, either to deftroy them, or fpoil their work.

To prevent their defign, and fecure the miners, a party of 10 or 12 grenadiers are ordered to follow them, with four or five fhells, fome with fufes only, and others loaded, the reft of the grenadiers carry grenades; and fo foon as the miners have worked their way into the Befieged's galleries, the grenadiers throw fome of the loaded fhells and grenades into them, and retire together with the miners. The effect of thefe fhells and the fmoak will, in all probability, prevent the enemies return for fome time; on the contrary, the befiegers return immediately and ftop up the hole with fand bags, leaving a fmall opening to throw more grenades in, in cafe the enemy fhould return; which if they happen to do before the hole is ftopped, the grenadiers throw in fome fhells with fufes only, the expectation of whofe burfting, or dread of whofe effects, will be fufficient to oblige the enemy to retire, whilft the Befiegers continue their work without interruption.

The miners ought to leave no means untried to difcover the enemies galleries behind the wall, which fometimes are not eafily to be come at, as well as to prevent their making any new ones, and to ufe the utmoft precautions not to be furprized themfelves, which the enemy will not fail to attempt, either to deftroy them or their works. The bufinefs of a miner requires great skill and conduct to avoid the ftratagems of his enemies; he fhould liften frequently to difcover whether the enemies are at work under or near him, and found now and then with his iron needle, or Augre, not only

towards

towards the place from whence he imagines any noiſe proceeds, but likewiſe all round him, taking particular care that the enemies, whilſt they make a noiſe one way on purpoſe to draw his attention there, do not fall upon him and ſurprize him by another, which is a common ſtratagem amongſt miners.

If the Beſiegers perceive that their enemies are coming toward them, a ſmall Mine muſt be made, to ſtifle them in their gallery, which may be done thus.

A hole of five or ſix inches diameter, and ſix or ſeven deep, is made of that ſide of the gallery where the enemy is expected, and ſome Powder made in a cartridge of the ſame ſize is put into it, which will contain about ten or twelve pounds; upon this powder is put a tampion, having a ſmall quick match running through it, to give fire to the powder, and this tampion is covered with ſtrong planks well buttreſſed; then as ſoon as the enemies are judged to be within four or five feet diſtance, this powder is fired, which will burſt their gallery, and either kill the miners, or the ſmoak will oblige them to retire.

Another way of burſting the enemies galleries is, by placing ſeveral ſhells in ſuch a manner as to have their effect that way. When the miners are in ſearch of each other, they pierce the interval betwixt them, in ſeveral places, with the iron needle, to find as near as they can their diſtance: The miner muſt watch for this opportunity, and ſo ſoon as the needle is withdrawn, clap a piſtol into the hole, which when well directed, ſeldom fails to kill the ſounder; the firſt ſhot may be followed by three or four more, and then the hole being cleaned,

and

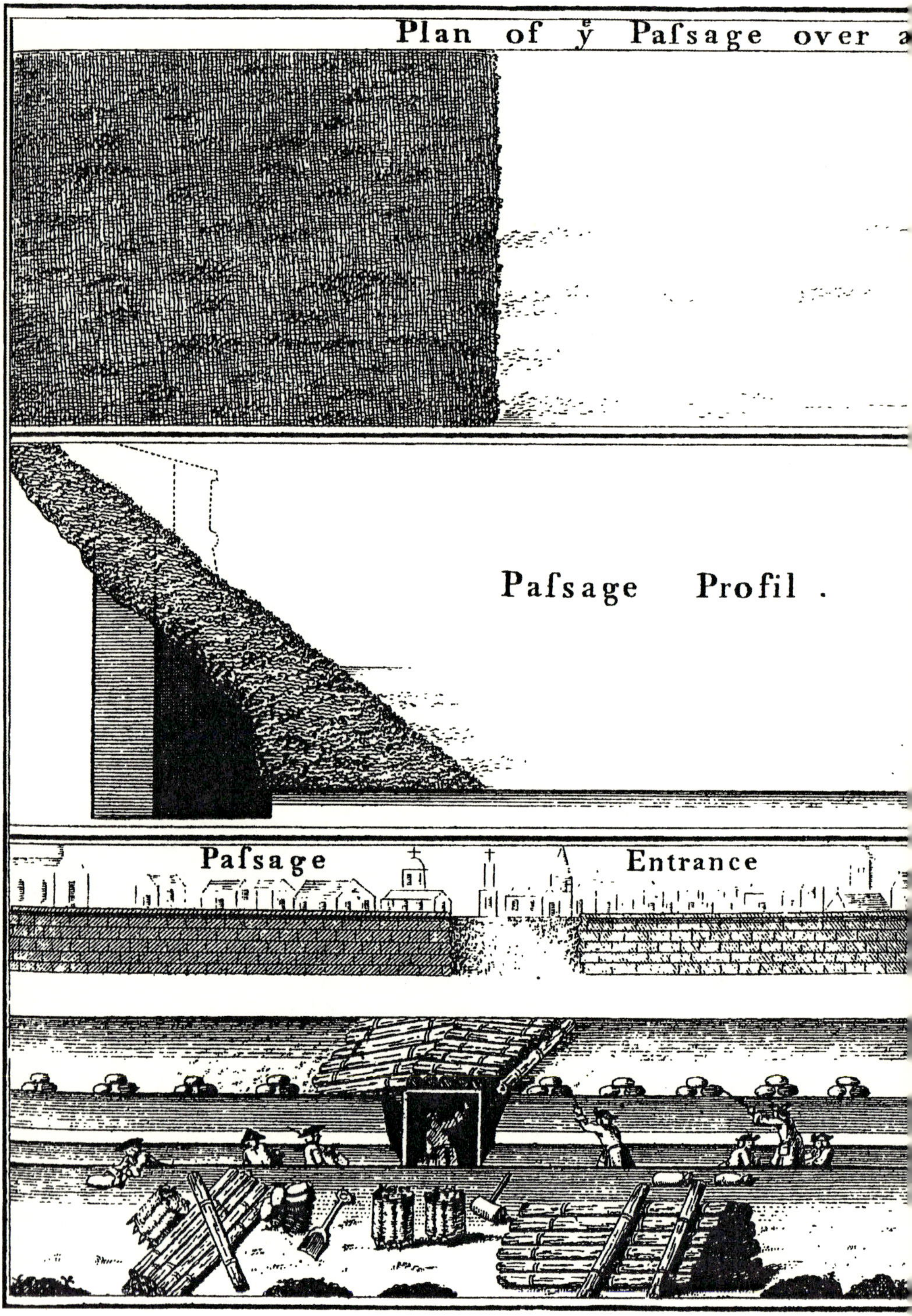
Plan of ye Pafsage over a
Pafsage Profil .
Pafsage
Entrance

'et Ditch .

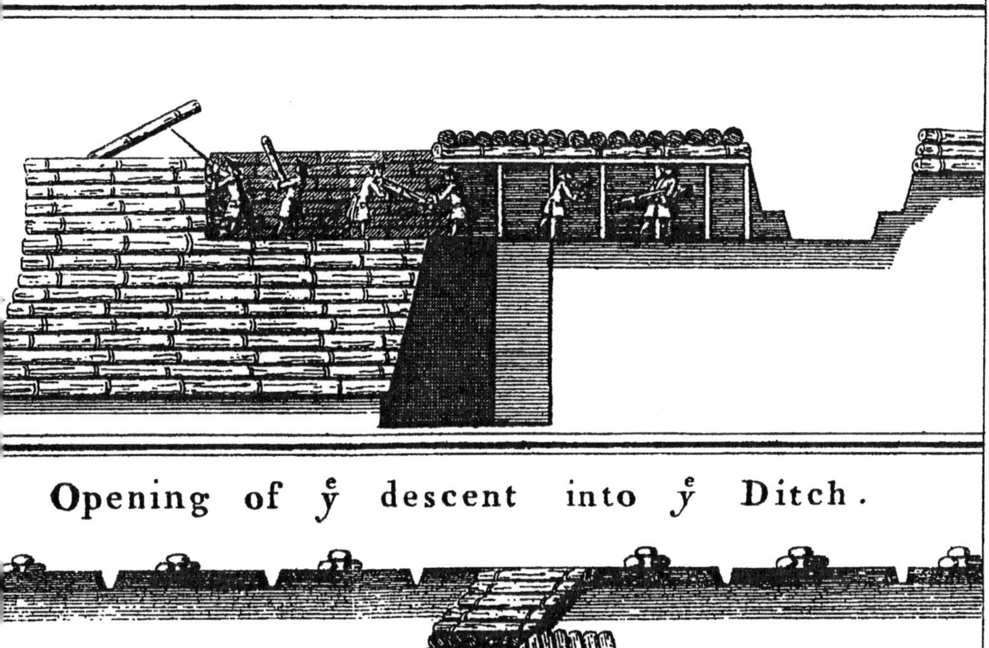

Opening of ye descent into ye Ditch .

Hulett Sculp.

and care taken that the enemies do not ſtop it on their ſide, a hollow cane filled with a ſtinking compoſition is introduced, and ſtopped on the inſide, which being fired, will make ſuch a ſtink as will force the enemy to quit their galleries.

Theſe and many other ſtratagems are uſed by skilful miners, not to be over-reached or ſurprized, whilſt the enemy are at the ſide or above them; but when they happen to be undermoſt, it is difficult to prevent being ſerved in the ſame manner.

RECAPITULATION.

WHAT has been ſaid with regard to the Attac of a regular Place fortified with ravelins and a covert-way, may, with proper caution, be applied to the Attac of any other Place likewiſe, whether regular or not, and fortified with any kind of out-works, obſerving for general maxims, *viz.* That the Approaches ſhould ever be ſo directed, as neither to be enfiladed by any work of the Fortification, nor yet to defile too much, which would occaſion more work than need; not to proceed uncovered after the fire of the Beſieged begins to be dangerous, but to carry on the works by ſaps; the firſt Parallel to be always within 300 fathoms or leſs of the ſaliant angles of the covert-way, if it may be done: the ſecond at 140, or thereabouts, from the firſt and the third, near or on the glacis; to make places of Arms, between the ſecond and third Parallels, at the turnings of the Approaches, that the troops placed there may be at hand to ſupport the workmen at the head of the Trenches, till ſuch time as the third Parallel is finiſhed; the miners to ſink ſhafts in or near the third Parallel, and carry on on their galleries under the works above ground and in the ſame direction, endeavouring to get under thoſe of the Beſieged, if poſſible, to prevent being blown up by them, and, at the ſame time, have an opportunity of blowing up theirs, whenever it ſhall be thought proper.

The firſt Batteries are placed between the firſt and ſecond Parallels, perpendicular to the faces produced of the works, and their number is always equal to that of the faces which flank the front attacked; from the third Parallel, double ſaps are to be made and carried directly over the ridges of the glacis, before the ſaliant angles of the covert-way; within 12 or 13 fathoms of theſe angles, cavaliers are made, nearly perpendicular to the oppoſite branches of the covert-way; ſaps are carried along the glacis towards the rentring angles, within three fathoms of the covert-way; which ſaps muſt be well traverſed to prevent the enfilades; and laſtly, Batteries are erected upon the ridge of the glacis, both to make breach, and to ruin the defences.

The fire from the cavaliers and *Ricochet* Batteries, both of guns and howitzes, and thoſe for throwing ſtones and ſhells, will together, in all probability, drive the Beſieged out of the covert-way, if not, they muſt be driven out ſword in hand: The Beſiegers being in poſſeſſion of the covert-way, Batteries are made in the gorges of the places of Arms to batter the oppoſite faces, curtains and tenailles, if there are any; two or three deſcents are made into the ſame ditch near one another; the paſſage over the ditch, the Breach and lodgment in it, are made according to the manner deſcribed before.

Theſe general maxims being well underſtood, a moderate capacity, and a little experience, will be ſufficient

ſufficient to underſtand Attacs of almoſt any kind of places whatever; but for a farther aſſiſtance I ſhall give ſeveral examples of places variouſly fortified and ſituated.

The foregoing method of attacking the ravelin firſt, and afterwards the baſtions, has been generally followed, eſpecially when the garriſon is ſtrong, and commanded by a skilful and reſolute Governor; but when this is not the caſe, both baſtions, and the ravelin before them, may be attacked together when the ditch is dry; for whilſt the breaches are making, and the flanks ruined, abundance of ſhells muſt be thrown into the baſtions, lower flanks, if there are any, and into the ravelin, to deſtroy the retrenchments in them; and making the deſcents into the ditches, either by ſaps or ſmall mines; when this is done a general attac may be made, without loſing any time, in making ſecure paſſages croſs the ditches, in the following manner: at dawn, ſtrong detachments of grenadiers, followed by others of volunteers, mount the three breaches at once, theſe muſt be ſupported by other troops; in the mean time a detachment of grenadiers enter by the gorge into the ravelin; and if there are caſematted flanks, they get between the embraſures, and fling ſome grenades into them, threatening the Beſieged to give them no quarters if they fire. If there is a ſally-port in the curtain between the baſtions attacked, a party breaks it open by means of a petard or otherwiſe, to enter likewiſe that way. This general attac being well concerted, and carried on with reſolution, will ſcarcely fail of ſucceſs, becauſe the Beſieged are ſurpriſed, not thinking of being attacked before the paſſage croſs the ditch is made in a regular manner.

The

The French carried Bergen-op-zoom by ſuch an attac in the late war, the garriſon not ſuſpecting in the leaſt to be in danger; they did not even think that the breaches were practicable.

The ATTAC *of a Place fortified with a* CROWN *or* HORNWORK.

PL. XIII. WE ſhall ſuppoſe, that the Approaches from the beginning to the third Parallel have been made in the manner already deſcribed, as being the ſame in all kinds of Attacs whatever; this plate repreſents that part only which is particular to a horn-work placed before a baſtion. L, L, L, being the third Parallel; a, a, a, oval traverſes, between the third Parallel and the direct double ſaps i, i, i, and b, b, b, the cavaliers of the trenches; c, c, Batteries before the places of Arms to throw ſtones; d, d, Batteries to make Breaches in the faces of the horn-work; e, e, Batteries for making a Breach in the ravelin; f, f, Batteries in the places of Arms, to fire at the faces, curtain and tenaille of the horn-works; g, g, Batteries for enfilading the covert-way before the branches of the horn-work, as likewiſe to fire at the faces of the oppoſite ravelins; k a lodgment in the Breach of the ravelin; l, l, lodgments in the Breaches of the faces of the horn-work.

The Trenches within the horn-work, have nothing particular from thoſe made in the ravelin; it may only be obſerved, that from the lodgments l, l, in the half baſtions, ſaps are carried on towards the curtain, and about half way, branches to the orillons to drive the Beſieged out of the tenailles; the firſt ſaps are continued quite round the curtain, before

before which a Battery m, m, is made, to fire upon the retrenchments within the horn-works, as likewiſe to annoy the fire of the baſtion; from the middle of the curtain a direct double ſap is carried on directly to the point of the baſtion; and from the extremities of the Battery m, m, other ſaps towards the parapet of the branches, and continued along them as far as the ditches before the retrenchments; theſe laſt ſaps are alſo join'd to the direct one r, by two others croſs the horn-work, ſo as that they may have a free communication with one another, and to place troops in the latter to ſupport the workmen: Laſtly, a Battery is made near the Counterſcarp of the great ditch to ruin the flanks which defend the baſtion P, and to make a Breach in that baſtion; then lodgments are made in the retrenchments within the horn-work, which are continued along their parapets to the counterſcarp of the great ditch.

All thoſe ſaps which are ſeen or enfiladed from any part of the place, muſt be well traverſed, and made ſufficiently deep, ſo as to be ſecure againſt the fire of the Beſieged.

The horn-work being taken, the two adjacent ravelins A, A, together with their redouts B, B, muſt be taken, before the paſſage over the ditch round the baſtion can be attempted, ſince theſe paſſages are ſeen from the ſaid ravelins: As to the reſt, the deſcent and paſſage over the ditch, the Breach in the baſtion, and the lodgments in it, are all performed in the ſame manner, and with the ſame precautions, as deſcribed before.

As there is no material difference between the attac of the horn-work and that of a crown-work; and whether theſe works are placed before a baſ-
tion

tion or curtain, the reader who underſtands the one will underſtand the other; ſo I ſhall inſiſt no farther on it: It may be obſerved, that a horn or crown-work placed before a baſtion, adds more ſtrength to the place than when placed before the curtain; in the firſt caſe, the two adjacent ravelins A, A, muſt be taken; and when the crown or horn-work is taken, there can but one baſtion be attacked, whereby the Beſieged are enabled to unite their whole ſtrength and defend it; whereas in the other, there is no occaſion to take any more than one ravelin, and two baſtions may be attacked together; by which the Beſieged will be obliged to divide their force, and conſequently are leſs able to reſiſt the Beſiegers.

The ATTAC *of a Place fortified with a ſecond* COVERT-WAY.

THE Approaches from the beginning ſo far as to the third Parallel, are carried on in the manner already deſcribed; PL. XIV. the ſecond covert-way is attacked in the ſame manner, and with the ſame precautions, as the former in Plate X. The lunets are likewiſe attacked as the ravelins, and with the ſame precautions; Batteries are raiſed in them to fire into the firſt covert-way, and when the Trenches are carried round their gorges, the ſeveral paſſages over the ſecond ditch are all made at the ſame time; but as the grazing fire from the firſt covert-way is very dangerous, great care muſt be taken to make the ſaps very deep, and to traverſe them well; as likewiſe to enfilade the covert-way with the ricochet-Batteries, and to throw

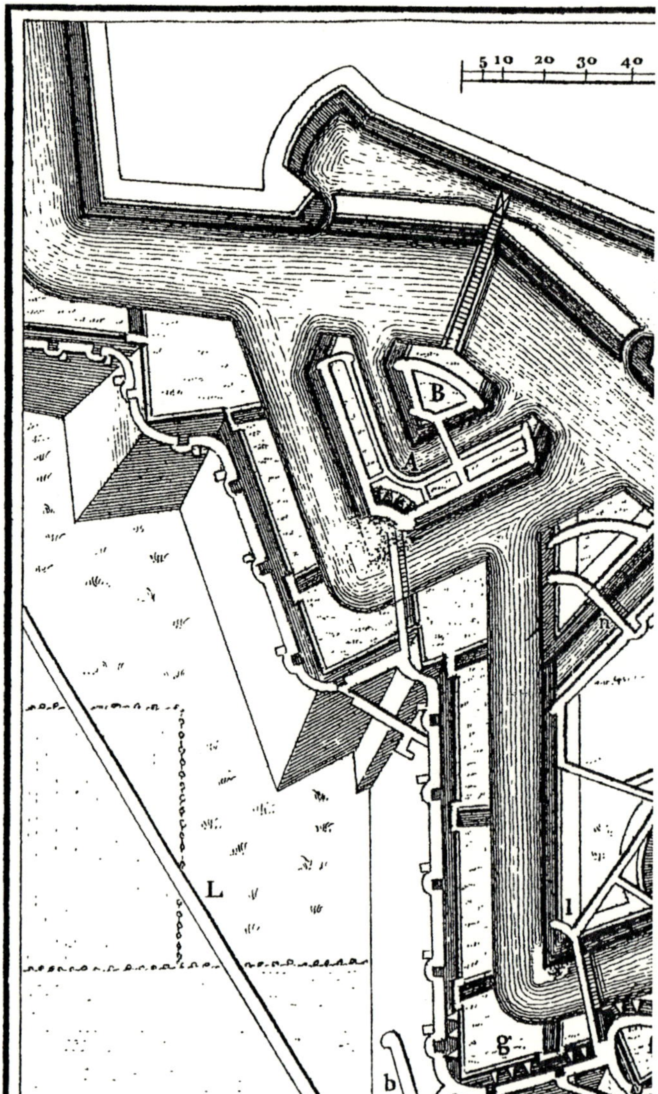
5 10 20 30 40
B
L
l
g
b
i
c

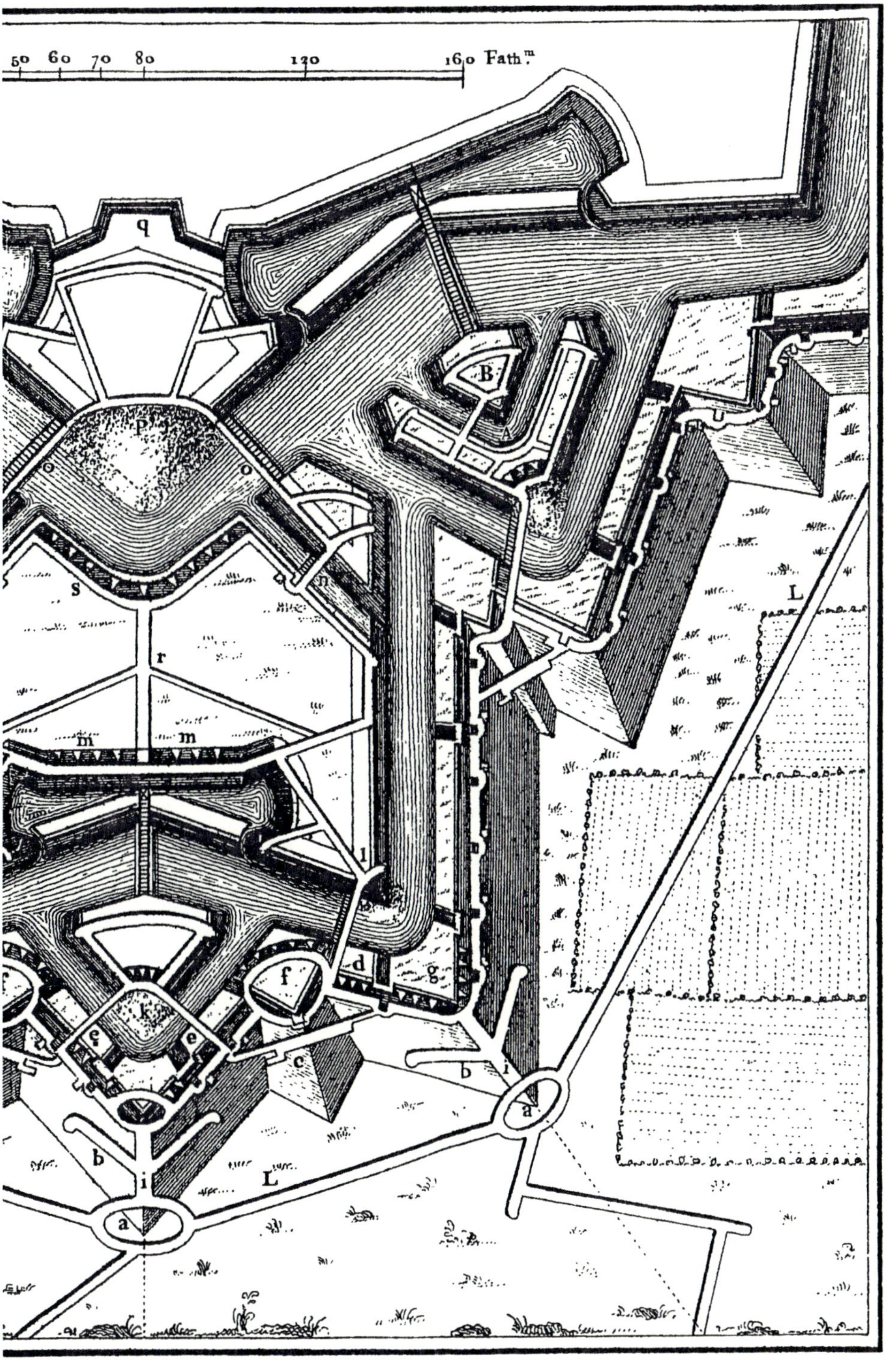

50 60 70 80 120 160 Fath.m
q
B
p
o
o
n
s
r
m
m
l
L
f
f
d
g
k
e
e
c
b
i
a
b
i
L
a

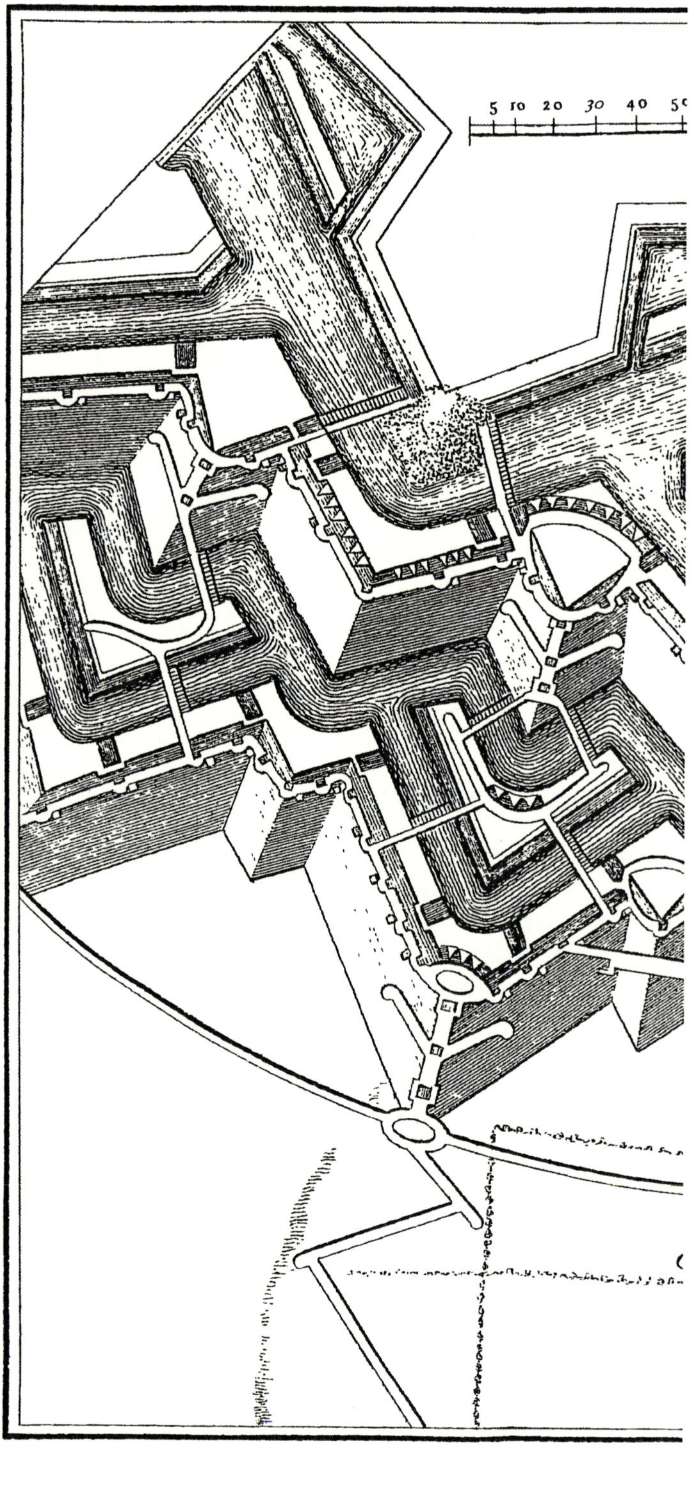
5 10 20 30 40 5

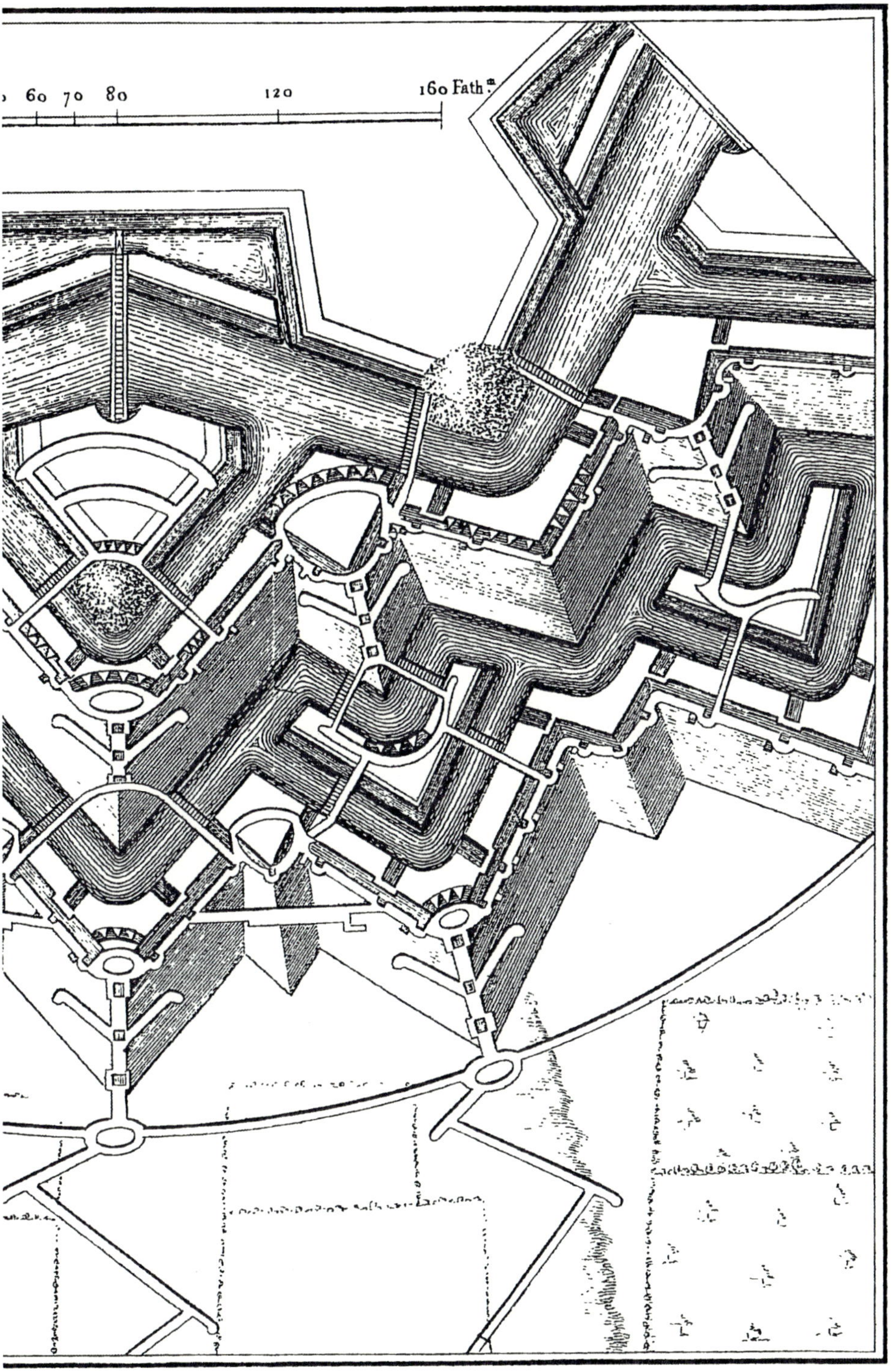
60 70 80
120
160 Fath.m

throw a great quantity of ſtones and ſhells into it at the ſame time, to diminiſh the enemy's fire as much as poſſible; without which, it would be impoſſible to make any lodgments on the firſt glacis.

This done, lodgments are made before all the ſaliant angles of the covert-way, and from thence ſaps are carried on directly over the ridges before theſe angles; about the midway, places of Arms or cavaliers are made at the right and left of theſe ſaps, as being abſolutely neceſſary both for protecting the works and to plunge into the covert-way, to abate its fire: At the ſaliant angles before the ravelin, an oval traverſe is made, ſo that the paſſage round it may take in the beſt part of that angle, and a Battery is made there, to annoy the enemy in the ravelin by an oblique fire; then the next traverſes muſt be taken, and the ſap carried on, on both ſides, along the glacis, as already deſcribed.

It is not to be expected, that theſe works can be made without great danger and blood-ſhed; for the Beſieged may ſally out when they pleaſe with all their forces, and fall ſuddenly on the workmen and their guards, drive them from their work, deſtroy them, and then retire with ſafety; whereas the Beſiegers can only oppoſe thoſe in the places of Arms or cavaliers; and as the latter cannot be ſo well ſuſtained by thoſe placed in the ſecond covert-way, on account of their great diſtance, and the intervention of the ſecond ditch, as to be in a condition to make head againſt them, the Beſieged have it in their power to do great miſchief. To prevent this ſo far as poſſible, the gun and mortar Batteries placed on the ſecond glacis, and in the lunets, muſt continually play into the covert-way,

as

as aforefaid, to enfilade it, and plough the ridge of the glacis in fuch a manner as that the Befieged dare not fhew themfelves; which, in time, will drive them quite out, and oblige them to retire into the places of Arms.

The lodgments on the glacis being advanced to the places of Arms, if the Befieged are obftinate in their defence, they muft then be driven out by a detachment of grenadiers, fword in hand, and lodgments made in them; as likewife Batteries in their gorges. This, and the Attacs of the ravelin and baftions are made in the manner already explained.

The writers on fortification feem not to have properly confider'd thefe kind of outworks; fo little notice has been taken of them, as would make one believe that they are but of little ftrength; but confidering the many Batteries neceffary to be raifed, all the feveral paffages over the fecond ditch, the number of lodgments to be made, with the great difficulty in maintaining them, on account of the defence thefe works mutually afford each other; I fay, all thefe well confider'd, are fufficient proofs of their excellency.

Mr. *Goulon,* tho' a great Engineer, as appears by his excellent memoirs on the Attac and Defence, feems not to have been fufficiently apprized of the advantages of thefe works; he confiders indeed the fecond ditch as dry, which very rarely happens, and when they are fo, they are made in fuch a manner as to afford no fhelter to the Befiegers, as he imagines.

On the contrary, this ditch is not only wet, but has generally fluices, which will make the water rife up half way the firft glacis; and in that cafe, how will the

the Befiegers make their trenches, to bring the earth from the fecond glacis, requires much time and trouble, and many gabions and fafcines; I fee no other remedy than to break thefe fluices with fhells, and as they are fecured with arches and cover'd with earth, a thirteen inch fhell will have no effect upon them; for which reafon the French have generally one or two eighteen inch mortars which they ufe upon fuch occafions, as they did at Aeth, where they fired three days before they could break the fluice that kept up the water in the great ditch.

The ATTAC *of a Place fortified with* DETACHED REDOUTS.

DETACHED Redouts, when they are well made, fhould be within musket-fhot of the covert-way, that is, within about 50 fathoms of it, that their communication may be well defended. But it happens fometimes, that they are made much farther, to occupy a fpot of ground which may otherwife be advantageous to the Befiegers; in which cafe, the Approaches are carried round fo as to take them by the gorge, whilft the mortar Batteries throw fhells and ftones in them; which will oblige the defenders either to retire betimes, or run the risk of being made Prifoners of War.

If the Redouts happen to be arched, as at *Luxembourg*, or their communications with the Place are under-ground, as at *Maeftricht:* In the firft cafe, the ftones and fhells have very little effect, if the fhells happen not to break them; the beft way then is, to make Mines under them and blow them up: And in the fecond, they may either be taken by carrying the Trenches round them, or

blow

blow them up likewiſe: The advantage of theſe latter kind is, that they may be defended to the laſt extremity, without expoſing the garriſon to be made priſoners, as in the others.

When they are within about fifty fathoms of the covert-way, a Breach muſt be made, and whilſt they are attacked, and lodgments made in them, the Batteries muſt fire ſmartly on all the parts of the covert-way, which defend them. But when the Redouts are quite cloſe to the glacis, they muſt be attacked at the ſame time as the covert-way; for if they are attacked before, the Beſieged will always be in a condition, on account of their nearneſs, to drive the Beſiegers out of them.

The Beſieged may make great uſe of Mines in the defence of ſuch works, whereby their Attacs would become very troubleſome; the only remedy in this caſe is, to make Mines alſo, and to get under the gallery of the Beſieged, if poſſible, or elſe it muſt be endeavoured to find them out, and either deſtroy their Mines, or render them uſeleſs.

If there are any Arrows at the ſaliant angles of the glacis they muſt be enfiladed by the ricochet-Batteries, and great quantities of ſhells and ſtones thrown in them, to oblige the defenders to retire.

ATTAC *of a Place fortified with* LUNETS, TENAILLIONS, *or* COUNTER-GUARDS.

APPROACHES, ſo far as the covert-way, being made as before deſcribed, the faces and branches of both tenaillons muſt be enfiladed by the ricochet-Batteries, as likewiſe the faces of the ravelin, if it can be done; then a Breach is made in each Tenaillon at the ſame time, and both at-tacked

tacked together, whilſt all the Artillery that can reach any part of their defence play continually upon them, as likewiſe the ricochet-Batteries, muſt endeavour to enfilade them every where: but if the Beſieged have made any traverſes to prevent the effect of the Ricochets, nothing but a great number of ſhells and ſhowers of ſtones will be able to annoy the Beſieged, and protect the Attacs, ſo as to enable the Beſiegers to make their lodgments in the upper part of the Breaches.

As the ravelin cannot be attacked till ſuch time the Tenaillons are taken, and the lodgments in them finiſhed; the Batteries which the Beſiegers are obliged to raiſe in the Tenaillons, to make a Breach in the ravelin, will be attended with great trouble and blood-ſhed, the ravelin defending theſe works wonderfully well on account of their nearneſs. The Beſieged will hardly give the Tenaillons up after the firſt taking them; but on the contrary, will return and drive out the Beſiegers in their turn, and hold them thus in play ſo long as they remain maſters of the ravelin, as it happened at *Liſle* in 1708, where the *French* drove the Allies out three ſeveral times. The condition of the Beſiegers will ſtill be worſe, if there are retrenchments in the Tenaillons, as there commonly are; for which no other remedy can be found, than to pour in ſuch quantities of ſhells and ſtones, that the Beſieged may have more need to ſhelter themſelves, than to think of obſtructing the Beſiegers in their work; and after the Tenaillons are taken, the retrenchments muſt likewiſe be taken, before the Beſiegers can be ſecure in the poſſeſſion of theſe works, and to be able to attac the ravelin.

If

If the ravelin has lunets adjoining it, but no bonnet before its ſaliant angle, a Breach may then be made in the faces of the ravelin at the ſame time as thoſe in the faces of both lunets, that all three may be attacked together: but if there is a bonnet to cover the ſaliant angle of the ravelin, it muſt be taken firſt; and when the lodgments in it are finiſh'd, Batteries are to be erected to make a Breach in the ravelins: but as this bonnet is well defended both by the lunets and ravelin, the maintaining theſe lodgments will meet with ſome difficulties till the reſt of the works are all taken: in the mean time the ricochet-Batteries muſt ſcower the lunets and ravelin, as alſo great quantities of ſhells and ſtones are to be thrown in them, to abate their fire as much as poſſible.

Theſe kind of works are capable of great Defence, as being ſo near each other, that the enemy may ſally out of the ravelin whenever they pleaſe, fall on the workmen in the lunets or tenaillons, drive them out, deſtroy their work, and then retire with ſafety; and this may be done ſo long as the ravelin is in their poſſeſſion: for which reaſon, they have always been eſteemed by the greateſt Engineers amongſt the beſt works that can be made, when the expence and nature of the ground will permit.

When a ravelin is covered by a Counter-guard, a Breach is made in the latter much in the ſame manner as in a ravelin; and, whilſt the Aſſault is making, a great quantity of ſhells and ſtones muſt be thrown into the ravelin, and the ricochet-Batteries try to enfilade its faces, endeavouring to drive the Beſieged from their defences, and to diminiſh their

fire;

fire; otherwiſe the taking the Counter-guard would be attended with great difficulties.

A lodgment being made in the Breach, ſaps are carried on along the parapets on both ſides towards the gorge, and Batteries erected to make a Breach in both faces of the ravelin; which, when made, as alſo the paſſage over the ditch, the ravelin is aſſaulted and taken in the manner deſcribed.

If there are Counter-guards before the baſtions, they are to be attacked at the ſame time as the ravelin; the narrowneſs of their rampart, and the vicinity of the baſtions, renders the making the lodgments in them very troubleſome; and as they are enfiladed by the oppoſite flanks, they muſt be made very deep and well traverſed; whilſt they are making, ſhells muſt be thrown into the flanks to abate their fire, otherwiſe there is no going on with the work. As theſe Counter-guards cover the flanks in ſuch a manner as not to be ſeen from the covert-way, they cannot be deſtroyed from any other place than from within them; and if they are double, and the Counter-guards narrow, the Batteries made in them will not be ſufficient to deſtroy the flanks; and therefore others muſt be made beſides theſe in the gorge of the ravelin, to batter them obliquely; but if the ravelin ſhould have a redout in it, it will be hard to find room enough for thoſe Batteries.

What has been ſaid in regard to the Attacs of Counter-guards placed before the Baſtions, ſufficiently ſhews their uſefulneſs, eſpecially when they are well made: but if their thickneſs is only twenty eight feet, as Mr. *Blondel*, and after him Mr. *Coehorn* will have it, there is not, they ſay, a ſufficient ſpace within them to erect Batteries for deſtroying the flanks;

flanks; whence the Befiegers will be obliged to bring earth from the covert-way to fill up part of the ditch, fo as to enable them to place a fufficient number of guns for ruining them; which muft confequently take up much time, and require great labour. But thefe Gentlemen did not confider, that three or four fmall Mines placed under thefe kind of Counter-guards will blow up fo much of them as to make an opening wide enough to batter the flanks from the oppofite covert-way; whereas, if inftead of twenty eight feet they are made about eight or nine fathoms broad, there will not even then be room enough to place a fufficient number of guns in them to deftroy the flanks, and yet they will be fecure from being blown up by Mines.

When the Counter-guards are nearly as high as the baftions, and have retrenchments lined with ftones or bricks, with a ditch before them, it is certain, that no kind of works whatever are able to make fo great a defence as they. The breach being defended by the faces of the baftions and the retrenchments, neither of which can be deftroyed before by cannon, makes it hardly poffible to get poffeffion of it and fecure a lodgment. All that can be done is endeavouring to deftroy the retrenchments by fhells, during the affault, to fling a great number of fhells into the baftions, flanks, and great fhowers of ftones into the retrenchments, to annoy the Befieged fo much as poffible. A lodgment being fecured, Batteries muft be erected both to make a Breach in the baftions and deftroy the flanks; which cannot be accomplifhed but with great labour and difficulty; and if the Befieged have mines to blow up thefe Batteries, they muft

be

be ſearched after and rendered uſeleſs, otherwiſe it would be impoſſible to maſter the place.

The French are ſo ſenſible that Counter-guards before the baſtions, are preferable to all other works, that they have built ſome before the baſtions of the citadel of *Liſle*, with retrenchments in them, tho' it had a double ditch and Covert-way: they did the ſame before the baſtions in the front that was attacked by the Allies in Queen *Anne*'s war, notwithſtanding that there are retrenched tenaillons before the ravelin.

The ATTAC *of a Place fortified with a* FAUSS-BRAY.

BY Fauſs-bray is meant a ſecond parapet going quite round the baſtions and curtains, nearly on a level with the inſide of the covert-way, and about four or five fathoms diſtant from the upper parapet. The interval between theſe parapets is a dry ditch, ſerving as a rampart to the Fauſs-bray; the Citadel of *Tournay*, and a few other Places are fortify'd after this manner: Mr. *Coehorn*, tho' he made uſe of them in his Treatiſe on Fortification, making the dry ditch from ſixteen to twenty fathoms broad, yet he never apply'd them, in any Place he has ſince fortify'd. Theſe Fauſs-brays being low, produce a grazing fire, which adds much to the ſtrength of the Place; but ſince *Vauban*'s Invention of Ricochet-firing, theſe low works are of very little uſe, there being nothing eaſier than to enfilade them from one end to the other; and as they are very narrow within, the ſhells and ſtones make a terrible havock amongſt the troops placed there; and if the upper rampart has

has a revetement, and that being fired upon, the rubbiſh which falls from thence will hardly permit the Beſieged to keep their poſts, much leſs to defend them.

The Beſieged being driven out of the Fauſs-bray by the ſhells, ſtones, and Ricochets, the paſſage over the ditch is made, as alſo the Breach in both parapets, the Rubbiſh of which will ſerve to fill up the dry ditch or ſpace between the parapets; if not, an epaulment muſt be made from one Breach to the other, at the ſide next to the Place; as likewiſe a lodgment may be made there to prevent the Beſieged from coming that way, and to ſupport the lodgment which is made afterwards in the upper Breach: as to the reſt of the works, they are made and carried on as uſual.

Explanation of PLATE XV.

A, B, Baſtions of the front of the attac.

C. Ravelin before that front.

D. Fauſs-bray.

b. Cavaliers of the Trenches.

c. Stone-batteries.

d. Batteries to make breach into the ravelin C.

e. Batteries againſt the defence of that work.

f. Paſſage over the ditch of the ravelin.

g. Lodgments in it.

h. Batteries againſt the defence of the Baſtions A, B, and that of their Fauſs-bray.

i. Batteries to make Breach into theſe Baſtions.

k. Batteries againſt the curtain.

l. Paſſages over the ditches before the Baſtions.

m. Lodgments in the baſtions and Fauſs-brays.

n. Paſſages made of faſcines to carry the guns over the Parallel to their Batteries.

The

P:91.

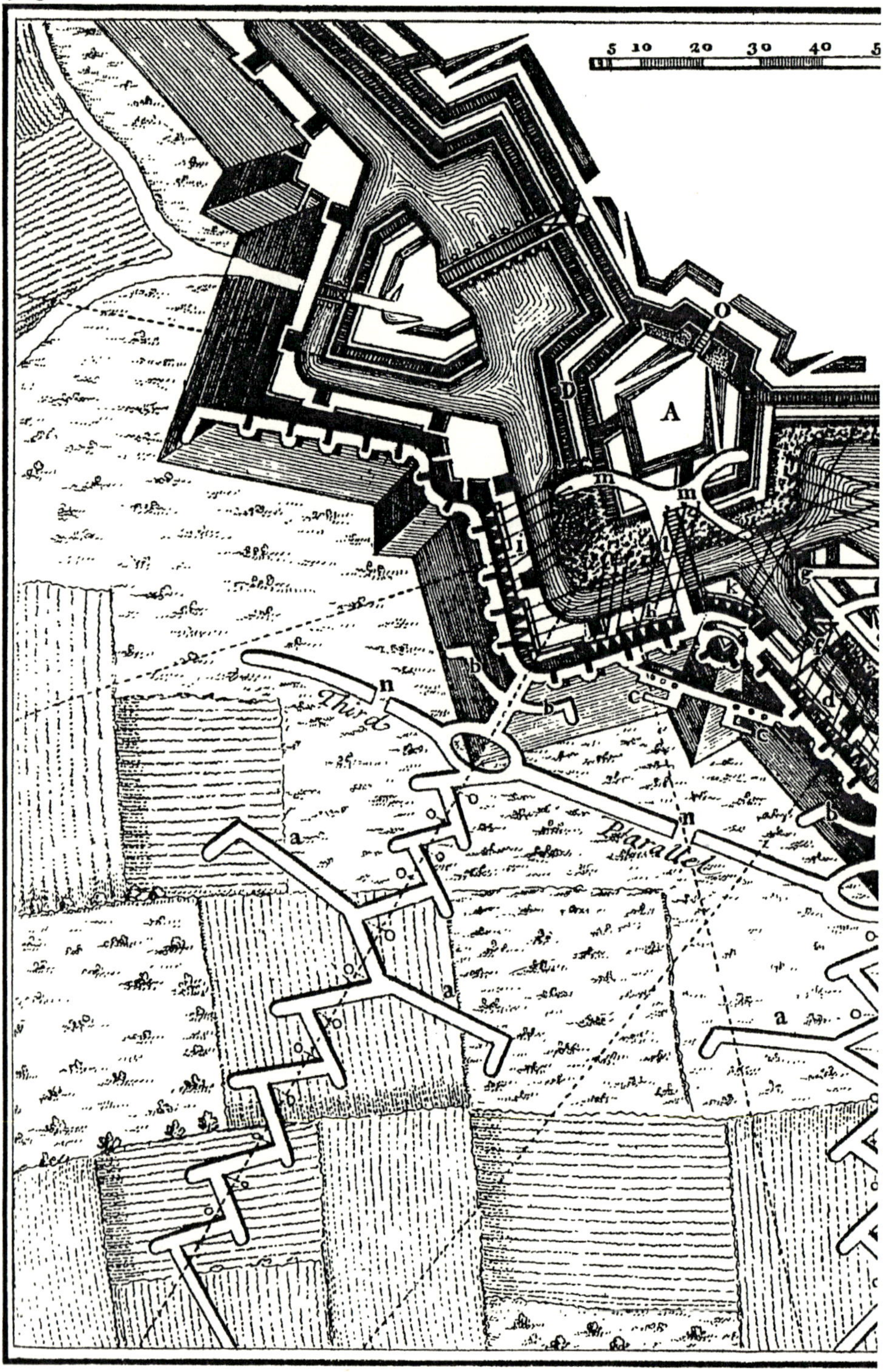

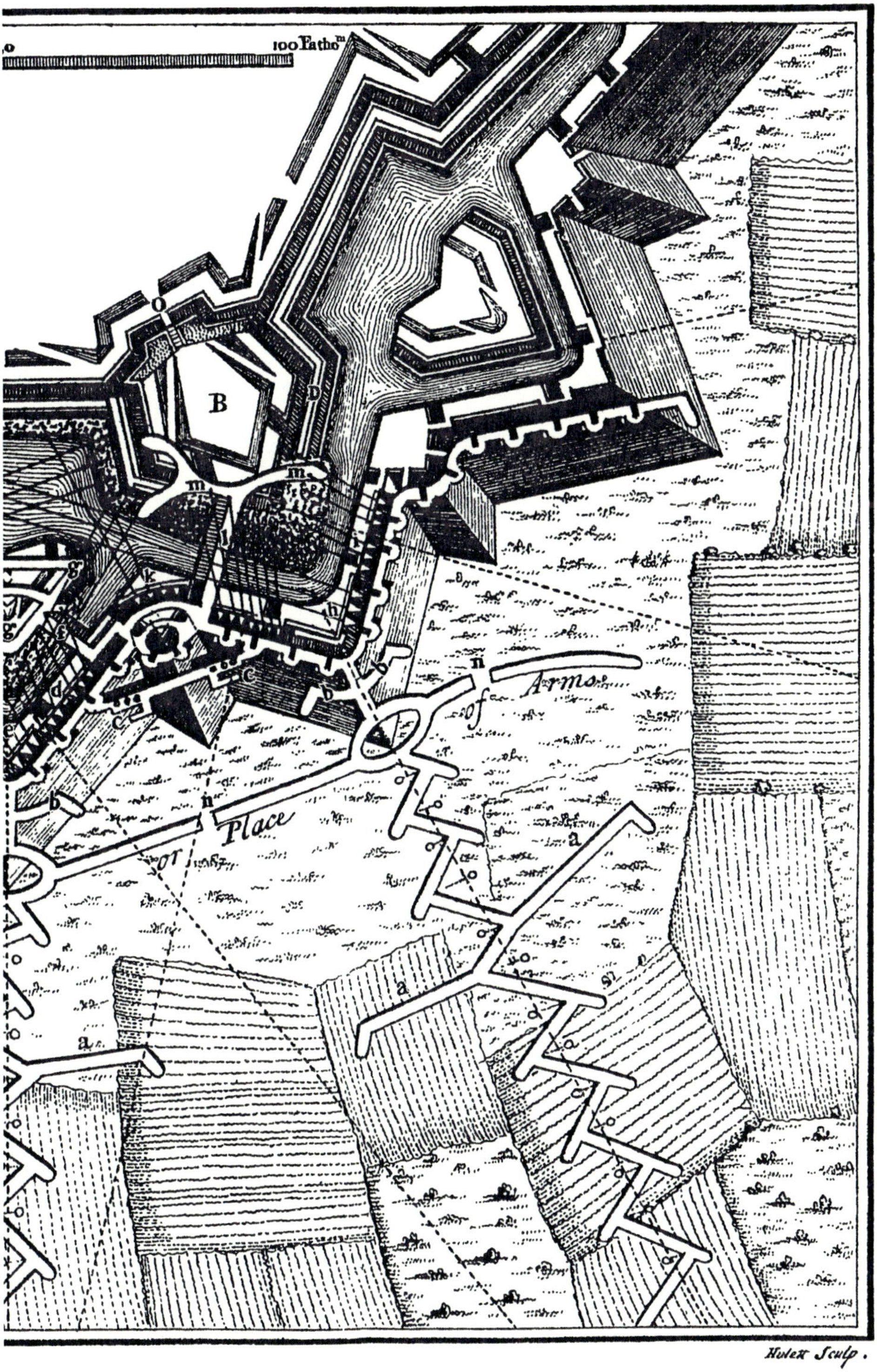

Hulett Sculp.

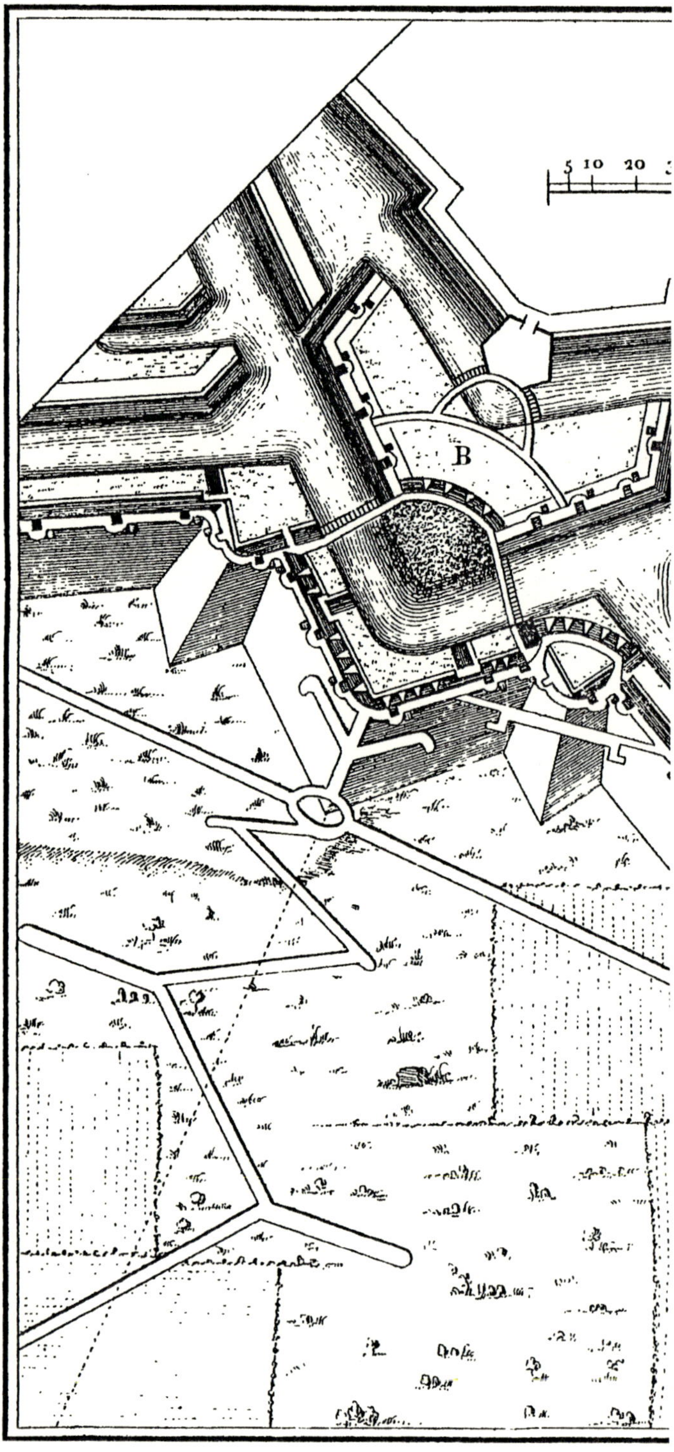
5 10 20
B

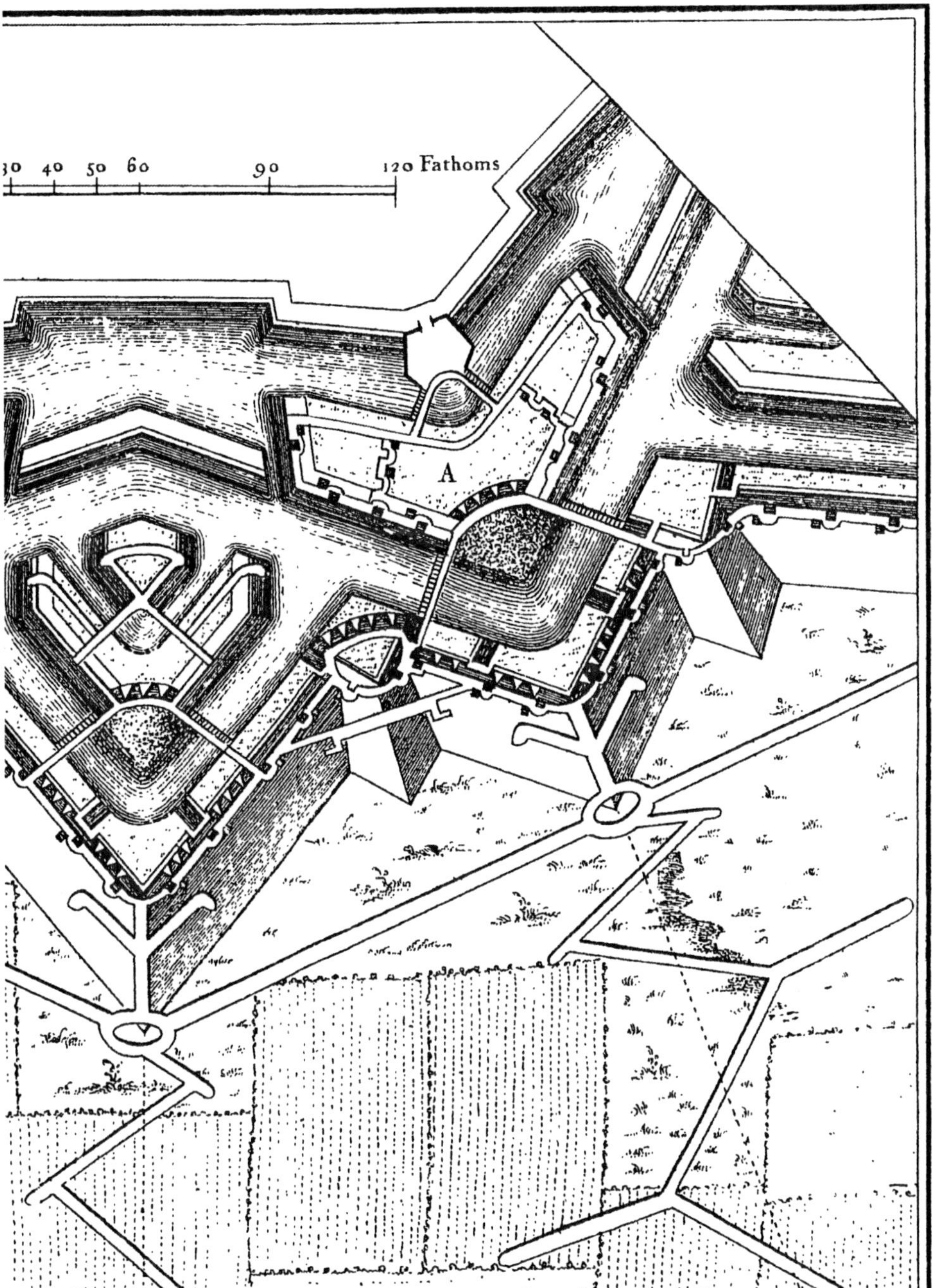
30 40 50 60 90 120 Fathoms
A

The ATTAC *of a Place fortified with* TOWER-BASTIONS.

PL. XVI. THE Trenches before ſuch a Place ſo far as the taking the covert-way, are in no reſpect different from thoſe of other Places: The counter-guards being nothing elſe but detached baſtions, they ſhould therefore be attacked in the ſame manner; but as they are ſeparated from the body of the place, they may be defended to the laſt extremity, without expoſing the Town to the hazard of being taken by ſtorm, and conſequently, they muſt be attacked with great caution; their defence muſt be thoroughly ruined, ſhells and ſtones thrown into them without intermiſſion, and their faces and flanks enfiladed by the ricochet-Batteries.

The Breach being made, and the lodgment in it finiſhed, a Battery is made at the ſaliant-angle, to batter the Tower-baſtion, and in the mean time ſaps are carried on along the parapets of the faces and flanks; from theſe lodgments near the flanks, the Beſieged are driven out of the tenailles, if they have not abandoned them before, and from the extremities of the flanks ſaps are carried on, on both ſides along the gorge, ſo as to meet before the point of the Tower. Theſe ſaps are joined afterwards to thoſe near the faces, by direct ſaps, ſo as to be able to ſupport one another, as may be ſeen in the counter-guard A; or the ſaps near the faces may be joined by a kind of round Parallel approaching the ditch before the point of the Tower-baſtion, without making any ſap along the gorge, as in the counter-guard B.

As

As the lodgment in the counter-guards are exposed to the fire of the Tower-bastions, and the ramparts adjoining to them, it will cost the Besiegers much time and numbers of Men, before they can erect Batteries to extinguish that fire, which, however, they will accomplish in time.

To come at the flanks of the Tower-bastions, the extremities of the flanks of the counter-guards which cover them, must be ruined by Batteries placed on the covert-way, which will not be easy to accomplish, as being about 24 feet in length, and the whole thickness of the parapet; after this the faces of the Towers must be battered to make Breach, at the same time the passage over the ditch is making.

When the passage is finished, and the Breach sufficiently wide, the Besiegers must endeavour to lodge themselves at the upper part of it; but as the Towers are casematted underneath, and these casemats divided into several apartments, the Besieged will be able to defend the Towers a great while; and till such time that the casemats are either stopped up or destroyed, it will hardly be possible for the Besiegers to lodge themselves in the Breach. Wherefore it will be necessary to make Batteries opposite to the faces, quite close to the ditch side, to batter the casemats; which being once destroyed, or stopped up, and the lodgments in the Breaches finished, the Besieged will hardly venture to hold out any longer.

Notwithstanding that the Tower-bastions may draw out the Siege to a considerable length, when they are defended as they ought to be, yet *Landau* was taken four times in the late wars, without attacking the Tower-bastions, the Place surrendering before-

before-hand; but whether this was owing to the imperfection of the works, the want of means to defend them, or the unskilfulneſs and timidity of the Commanders, is a queſtion undecided.

This method of attacking a Place fortified with Tower-baſtions, is that which its inventor imagined would be perſued; but if inſtead of attacking the Tower-baſtions themſelves, the adjacent ramparts were attacked, it would, in my opinion, make the work much eaſier; for the flanks and faces of the Towers being well battered, to render the guns in the caſemats uſeleſs, a Breach might be made in the rampart, without much difficulty; then all the great defence which *Vauban* propoſed with theſe under-ground works can be no obſtacle.

It is well known, that in a ſhort time theſe caſemats are ſo filled with ſmoak, that there is no ſtaying in them, notwithſtanding their chimneys, and as theſe Towers are made of ſolid maſonry, they are quite rejected at preſent by the beſt Engineers, as not anſwering in the leaſt the expence of building them.

The ATTAC *of a Place fortified with* RAMPARTS *without* BASTIONS, *according to the ancients, and covered with* OUT-WORKS, *in the modern manner.*

WHEN ſuch Places are fortified with counter-guards and ravelins, as cuſtomary, they differ not much from a Place fortified with Tower-baſtions; they generally have large wet ditches, filled with running water, as at *Douay, Tournay, Barcelona,* &c. The out-works of theſe Places are attacked in the ſame manner as uſual; but as the wideneſs of the ditch renders their communication with

with the body of the Place inconvenient, and eafily to be cut off, they cannot make all the refiftance which they otherwife might, if their retreat was more fecure; it is for this reafon that the Befiegers erect Batteries in fuch places as have a view of the communications, which foon obliges the Befieged either to quit betimes their out-works entirely, or to leave but few men to defend them, who are generally made prifoners of War.

If the ditch before the body of the Place is full of running water, the paffing it requires * much time and trouble, on account of its largenefs; two or three paffages muft be made, and as many Breaches, which the Befieged may defend for fome time, efpecially when the garrifon is pretty ftrong; but the fuperiority of the Befiegers both in men and Artillery, will get the advantage over the Befieged, let them be ever fo obftinate, and therefore muft at laft fubmit.

PLATE XVII. *Explained.*

A, B. Towers of the old rampart.
D, E, F. outworks, which cover thofe towers.
a. Half Parallels, or places of Arms.
b. Cavaliers of the Trenches.
c. Stone Batteries.
d. Batteries to make Breach in the pieces D, E, F.
e. Batteries againft the defences of thefe pieces.
f. Paffages over the ditches before them.
g. Lodgments in them.

h. Batteries

* The Allies attacked *Douay* in the late Wars, near the gate of *Notre-Dame*, and tho' there were but few out-works, yet the Town made a vigorous Defence for fix weeks. This Place has fince been well fortified by the *French*.

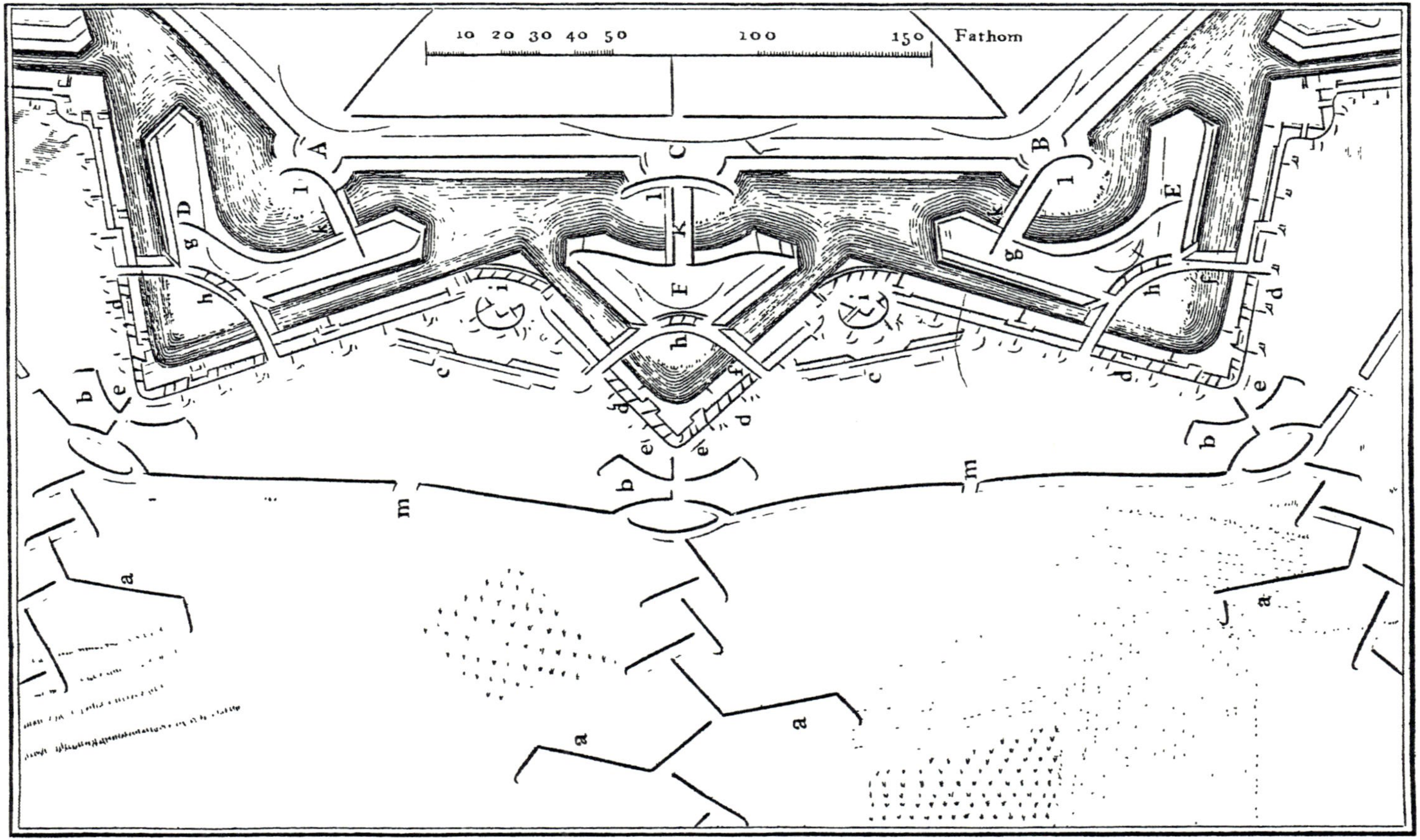
10 20 30 40 50 100 150 Fathom
A
B
C
D
E
F
K

h. Batteries to make Breach in the Towers A, B, C.
i. Batteries againſt the curtains.
k. Paſſages over the ditch before the Towers A,B,C.
l. Lodgments in theſe Towers.
m. Paſſages of faſcines in the third Parallel to carry the guns over to their Batteries.

ATTAC *of a Place fortified with* CAVALIERS.

THE Cavaliers are generally placed within the baſtions, and made of the ſame form, leaving about 5 fathoms room between the parapets, ſerving as a rampart to the baſtion, and their height exceeds that of the baſtion by about ſix or ſeven feet; ſome times they are placed elſewhere, as in the gorges, or behind the curtain; but this happens ſeldom.

Cavaliers are only made when the Town is commanded by ſome hill or riſing ground within cannon-ſhot, as at *Aeth* and *St. Omer*; they oblige the Beſiegers to open the trenches far off, and when the Approaches are advanced near the glacis, the Cavaliers plunge into them, which, on that account, muſt be made deeper than uſual; Batteries are made to enfilade them, if poſſible, otherwiſe a great number of ſhells are thrown into them, endeavouring to diſmount their guns; but as the throwing ſhells is pretty uncertain, and the Batteries of cannon can do them no great damage when they are near, and not able to reach them when far off, it ſeldom happens that all the guns can be diſmounted, as it happened at *St. Omer* in Queen *Ann*'s War, where two guns placed in a Cavalier could never be diſmounted by the Allies, during the whole Siege; for which reaſon the Batteries of mortars muſt continually

tinually play upon them, to leſſen, at leaſt, their fire, if the guns cannot be diſmounted.

When the miners get under the baſtion, they muſt carry their galleries under the Cavaliers likewiſe, to make Breach in both at the ſame time; but if the Cavaliers are placed near the gorge of the baſtion, as done by ſome Engineers, the baſtion muſt be taken, and lodgments made in it, before the Cavalier can be attacked. If there is a ditch before the Cavalier, with a row of paliſſades in the middle, they muſt be deſtroyed with guns placed in the baſtion, or elſe the miners muſt get under them to blow them up. As the inſide of the baſtion may poſſibly be full of mines, the Beſiegers muſt act with great caution, ſo as not to be blown up at every ſtep they advance: it will therefore be proper to carry galleries under the works above-ground, to ſecure themſelves, eſpecially when there is any ſuſpicion of Counter-mines, which may be gueſs'd at by the proceedings of the Beſieged, the capacity and reſolution of the Governor, and ſtrength of the Garriſon.

If all the Places were ſituated on even ground, the general rules already given would afford ſufficient directions for carrying their Attacs; but as moſt of them are ſurrounded either by rivers, brooks, or marſhes, or are ſituated on hills or riſing grounds, ſo as to be only acceſſible in ſome parts, the direction of the Trenches varies accordingly; and therefore chiefly depends on the ſkill and experience of the Engineers, who are to take ſuch meaſures as are moſt ſuitable to their ſituations. All the different circumſtances which they are attended with are impoſſible here to be particularized; for which reaſon I ſhall make ſome general obſervations,

tions, to give at leaſt ſome idea of what ought chiefly to be attended to in the Attacs of Places variouſly ſituated.

The ATTAC *of a Place ſituated on uneven ground.*

IF the Place be ſurrounded with a moraſs or low ground, ſo as only to be approached by ſome cauſeway, the firſt thing to be done is, to try whether the marſhes may not be drained, and the water turned into ſome other channel, which may ſometimes happen; but if that is not to be done, the Trenches muſt be carried along the cauſeway, widening them as much as poſſible, and making Batteries in the moſt convenient places, ſuch as are to be found, or may be made for that purpoſe; and as the ground will ſcarcely allow of making Parallels, or places of Arms, half Parallels, muſt be made at proper diſtances, to protect the workmen in the beſt manner that the nature of the ground will admit.

If the cauſeways happen to be ſo little above the level of the water as not to furniſh a ſufficient quantity of earth for making the Approaches, they muſt be made with gabions, faſcines and ſand-bags; and ſometimes the Trenches are made wider than ordinary, to furniſh a ſufficient quantity of earth for raiſing their parapets. It muſt be obſerved particularly, whether the cauſeways are enfiladed wholly, or in part only, by the works of the Place; if they are every where enfiladed, and there is no opportunity of turning the Approaches, ſo as to avoid the greateſt part of the inconveniency, the Attac will, in ſuch caſes, be next to impoſſible; and if there are no places to be found for erecting Batte-

ries,

ries, to annoy and enfilade the works of the Place, the condition of the Befiegers will ftill be worfe.

If, notwithftanding all thefe obftacles, the Siege of fuch a Place be refolved upon, then the foundations where they are bad, muft be mended with fafcines or hurdles, covered with fome earth brought from fome other place, and the parapets made with gabions and fand bags, and the work carried on in a direct double fap frequently traverfed. The reader will imagine that work of this nature is not to be carried on a great way; we fuppofe that there is not above 50 or 60 fathoms at moft that requires it.

It is in a fituation like this, that an Engineer may difplay his ability, fince there can hardly be a more difficult one to be found; but fhould all his endeavours prove vain and fruitlefs, the beft and only way left is to reduce fuch a Place by famine, as being eafily blocked up by a fmall number of troops, they can feldom hold out long, on account of their unhealthy fituation, and oft are fhort of provifion.

But if the caufeways are fix or feven fathoms broad, and five or fix feet above the level of the water, and if any places are befides to be found, or may be made with little trouble, for erecting Batteries, fo as to enfilade the works of the Place; fuch an Attac may be made, tho' not quite fo conveniently as if it were fituated in an open country.

Amongft the difficulties to be encountered in attacking a Place fo fituated, there is this advantage, that as the country muft be unwholefome, the garrifon is commonly fickly, and not in a condition to undergo the fatigues of a long Siege; befides, few Places are fo entirely furrounded with marfhes,

but

but that ſome part or other about them may be proper for making Approaches; it is indeed certain, that thoſe parts, which are eaſieſt of acceſs, are always better fortified than the reſt, however it is beſt to make the Attacs on that ſide; for the advantages of the ground will be more than equivalent to the additional ſtrength of the Fortification. But it will not be improper to try whether the marſhes are not to be paſſed over by ſome means or other. If on the one ſide the difficulty of paſſing be compared with the ſurplus of the ſtrength of the works on the other, one may form a judgment, which of the two Attacs will be eaſieſt.

When the *French* beſieged *Philipsbourg* in 1734, had they known that the marſhes extended only about 50 yards, as they found afterwards, when the Town had ſurrendered, they might have been much ſooner maſters of the Place, and would have ſaved great expence and many lives; which evidently ſhews the importance of thoroughly examining the ground about the Place, before the Attac is reſolved upon; for every thing tho' trivial in appearance may contribute to advance or retard a Siege conſiderably.

At the ſaid Siege of *Douay*, General *Armſtrong* finding that a cauſeway kept up the inundation which the *French* had made on the ſide near the Attac, propoſed to the Duke of *Marlborough* the draining of the water, by means of a trench a-croſs this cauſeway; which being done, he ſucceeded to his wiſhes, for it greatly helped to forward the taking of the Place.

Explanation of PLATE XVIII.

A, B, C, D. Towers, which flank and form the front attacked.

E, F, G.

E, F, G. The out-works which cover theſe Towers.
H. A ſecond ditch at the foot of the glacis.
I. Cauſeways, by which the Place can only be approached.
K. Trenches made on theſe cauſeways.
L. Ricochet-Batteries, to enfilade the faces and covert-way of the work E.
M. Ricochet-Batteries, to enfilade the faces and covert-way of the work F.
N. Ricochet-Batteries, to enfilade the faces and covert-way of the work G.
O. Mortar Batteries.
P. Trenches round the ſecond ditch.
Q. Paſſages over the ſecond ditch.
R. Cavaliers of the Trenches, which enfilade the covert-way.
S. Stone Batteries.
T. Trenches upon the ridge of the glacis.
U. Batteries to make Breaches in the works E, F, G.
X. Batteries againſt the defences of } theſe works.
Y. Paſſages over the ditches before } theſe works.
Z. Lodgments in } theſe works.
a. Batteries to make Breach in Towers A, B, C.
b. Batteries againſt the curtains.
c. Paſſages over the ditches before theſe Towers.
d. Lodgments in the ſaid Towers.

N. B. As the ground is ſuch that no ſallies can be made, the half Parallels, or places of Arms, are omitted here.

The ATTAC *of a Place ſituated near a great river.*

PLACES near rivers are not ſo difficult of acces as thoſe ſurrounded by marſhes; their Approaches

P:101.

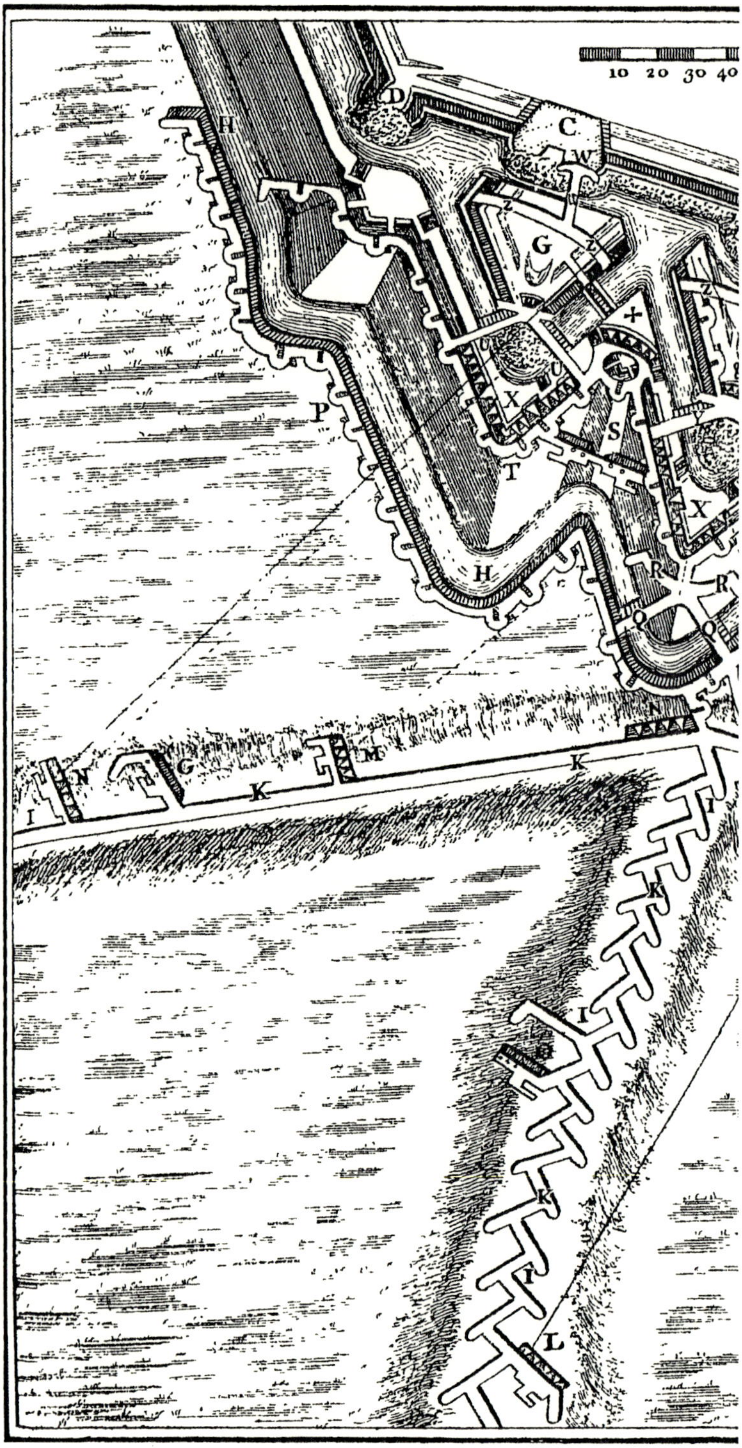

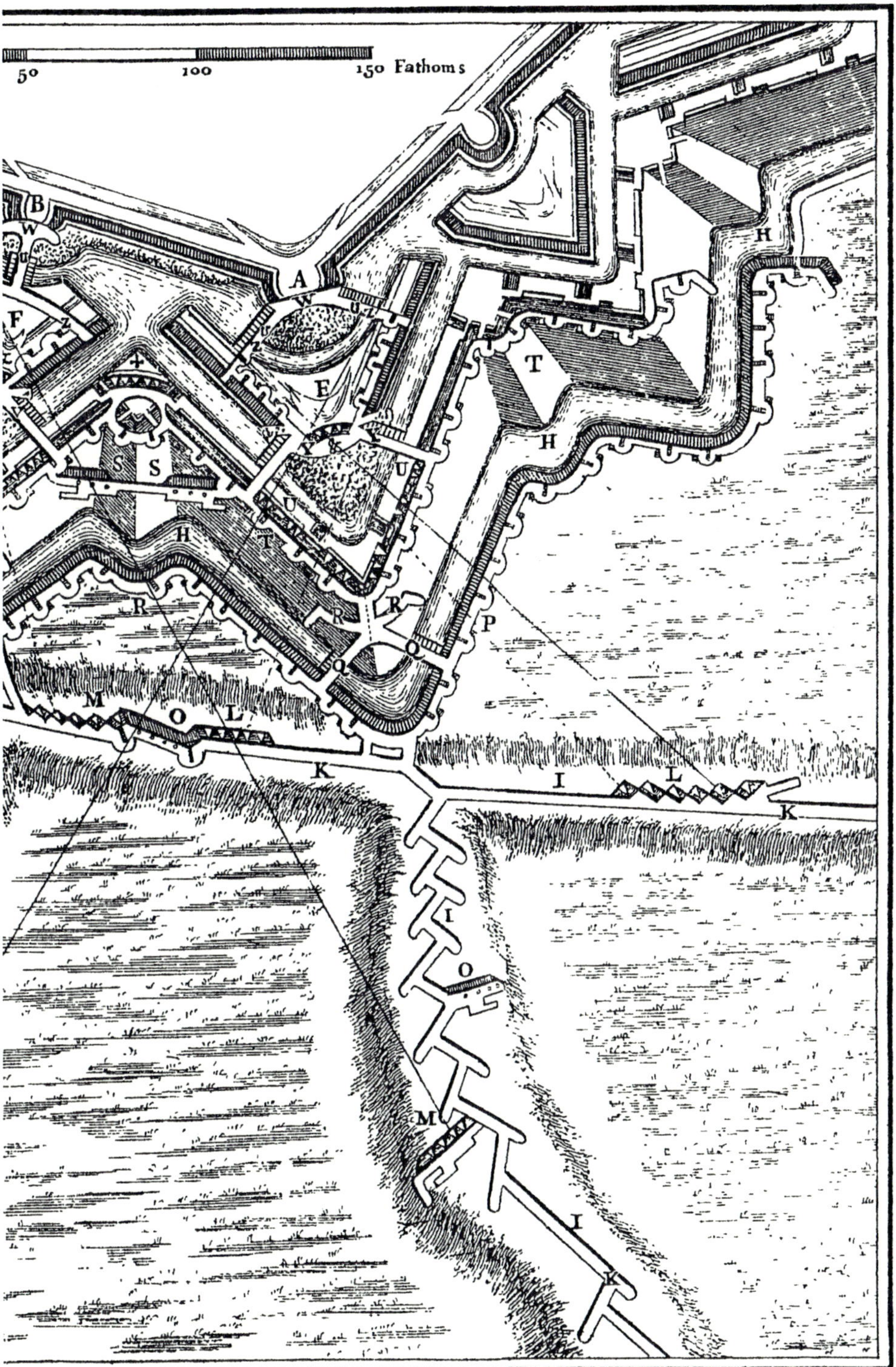

Hulett Sculp.

proaches are made as ufual, on the fide moft convenient; which generally is that next the river, if it is large, and does not run through the Town; but, on the contrary, when it does, the Attac is then made on the fartheft fide from the river, as before mentioned.

If the Attac is made on the river fide, Batteries are raifed on the oppofite fhore, to ruin the defences of the works which are feen from thence; as alfo to enfilade, and even to make fometimes a Breach from that fide. *Vauban*, at the Siege of old *Brifac* in 1703, made a Battery in an Ifland formed by the *Rhine* to batter the works attacked, and to make Breach, by which the Siege was much accelerated; and at the Siege of *Fort Keil*, in 1733, Batteries were erected in an Ifland, which made Breach in the horn-work attacked, as likewife in the baftion behind the horn-work; the fame Batteries enfiladed befides the covert-way, before that baftion that defended the branch of the horn-work next the *Rhine*.

At the above-mentioned Siege of *Philipsbourg*, the *French* took Fort St. *Michael* on the other fide of the *Rhine*, and erected Batteries there, which enfiladed the defences of the front attacked, and thereby prevented the fire of the Befieged from much difturbing their Trenches.

When there is a bridge over the river, it is generally covered by fome Fort, as at *Hunninghen*, *Philipsbourg*, &c. in that cafe Batteries are placed near the river, fo as to deftroy that bridge before the Fort is attacked; which being effected, the Fort will foon be obliged to furrender, or elfe the garrifon will be made prifoners of war.

It

It is an obſervation of importance, to know at what time the river overflows, and the extent of its inundation, to ſecure the Trenches from being overflowed, as likewiſe to place the Batteries and the park of Artillery ſo as not to be incommoded thereby. It often happens that rivers become fordable along the ſhore in dry weather, or when the waters are low, of which ſuch uſe may be made, by a good Engineer, as to ſhorten the Siege conſiderably.

The *Rhine* has that particular quality, of always overflowing in the hotteſt ſeaſon of the year, by the melting of ſnows on the mountains in *Switzerland*.

As there are few Places but have rivers running by or through them, convenient for traffick, it is the buſineſs of a chief director to draw all advantages from ſuch a ſituation, to carry on his Attacs; ſometimes when it is but middling, the Attacs are made on both ſides, for the conveniency of bringing the Artillery, ſtore and ammunition by water near the place where they are wanted, as Prince *Eugene* did at the Siege of *Liſle* in 1708; tho' he knew very well that the front which he attacked was not the weakeſt. Many other advantages may accrue from ſuch ſituations, which are impoſſible either to be foreſeen or deſcribed; and only obſerved by an able Engineer, who is upon the ſpot.

Explanation of PLATE XIX.

A. Cavaliers of the Trenches.

B. Stone Batteries.

C. Batteries to make breach in the ravelin before the horn-work.

D. Bat-

D. Batteries againſt the defences of the ravelin.
E. Paſſages over the ditch before that ravelin.
F. Lodgments in it.
G. Batteries againſt the flanks of the horn-work.
H. Batteries to make breach in the faces of the horn-work.
I. Batteries againſt its curtain.
K. Paſſages over the ditch before the half baſtions.
L. Lodgments in theſe half baſtions.
M. Batteries to make breach in the ravelin 5.
N. Batteries againſt the defences of this ravelin.
O. Paſſages over the ditch before the ravelin.
P. Lodgments in it.
Q. Batteries againſt the curtain of the body of the Place.
R. Batteries againſt the defences of the baſtions 6 and 7.
S. Batteries to make breach in theſe baſtions.
T. Paſſages over the dry ditches.
U. Lodgments in the ſaid baſtions.
X. Roads to carry the guns and mortars to their Batteries.
Y. Paſſages of faſcines in the third Parallel to carry the guns and mortars to their Batteries.
Z. Batteries on both ſides of the river.

The ATTAC *of a Place ſituated on a Hill.*

WHEN a Place is ſituated on a hill, having only a narrow ſpot of ground before it to make the Approaches, without any convenient place for erecting the Batteries to enfilade the works; the Attac is very troubleſome: it muſt be obſerved whether there is not ſome hill or riſing-ground near enough to make Batteries, ſo as to plunge or enfilade

filade ſome part of the Fortification, or if there be any from whence ſome works may be ſeen tho' obliquely: but if no ſuch places are to be found, the works muſt be battered in front, and mortar Batteries muſt likewiſe be made to ply them well with ſhells, to give the Beſieged all the diſturbance poſſible.

As to the diſpoſition of the Parallels and Approaches, they muſt be made and directed as the nature of the ground will allow, to anſwer as well as poſſible the deſired effect.

It may be obſerved, that very high Places, when they can only be battered from low grounds, are partly out of reach of the ricochet-Batteries; becauſe when a gun lies with its breech upon the bed, without any coins underneath it, it ſeldom riſes its direction, above twelve degrees; which range will hardly reach any work above fifteen fathoms high.

When Approaches are to be carried from a low-ground to a Place ſituated on a high-hill; all poſſible expedition muſt be uſed to get near the works of the Place ſo ſoon as can be; and the Trenches muſt be made very deep and narrow. For ſuch a ſituation give the Beſieged not only an opportunity of plunging into the Trenches, but likewiſe of ſallying with great advantage, which renders the ſupporting the trenches very difficult and dangerous.

Places ſituated on hills have generally no other but rain-water within their walls, ſo it muſt needs be ſcarce in dry weather; for which reaſon every poſſible precaution ſhould be taken to ſecure all the ſprings and brooks which are near the Place; by which

P:104.

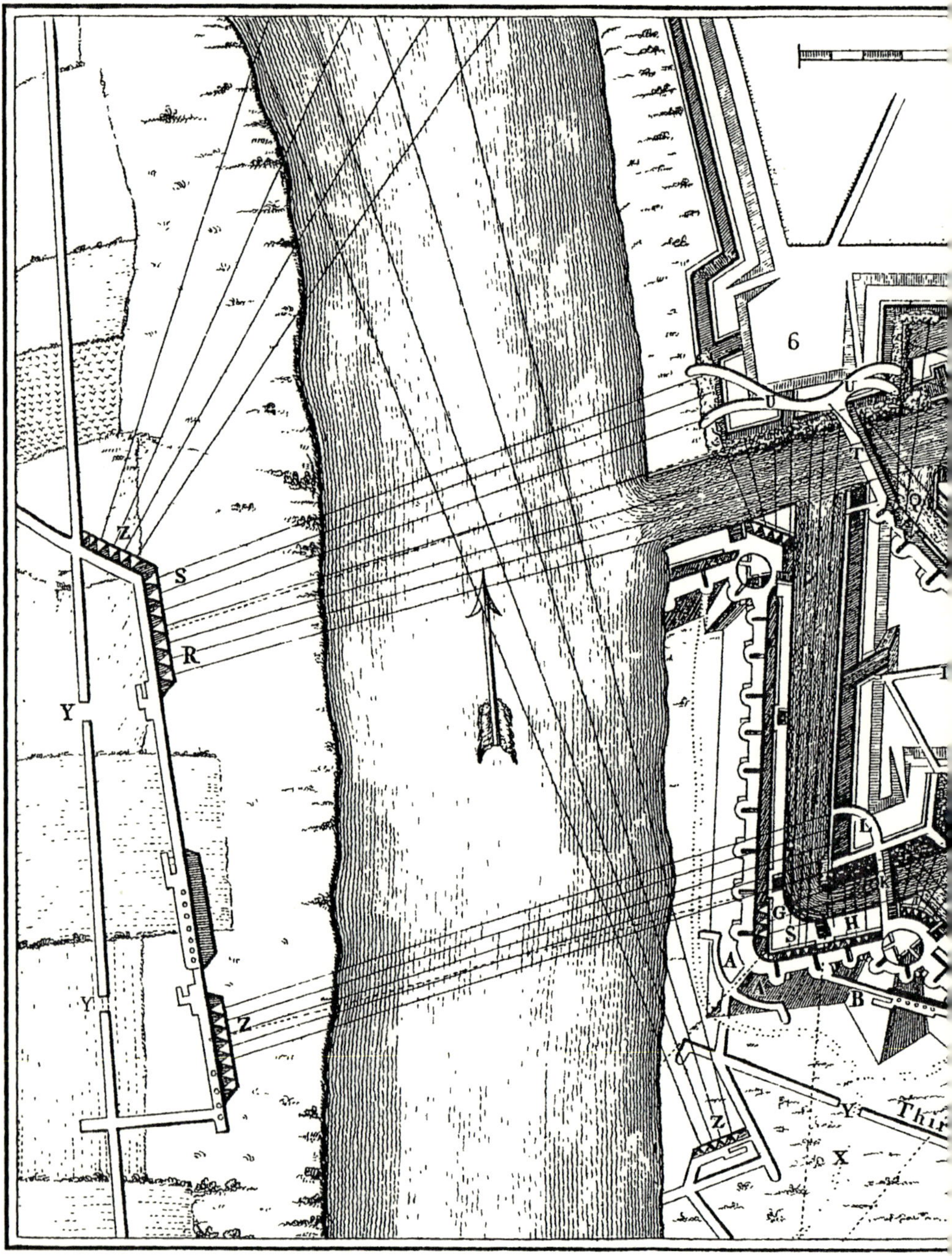

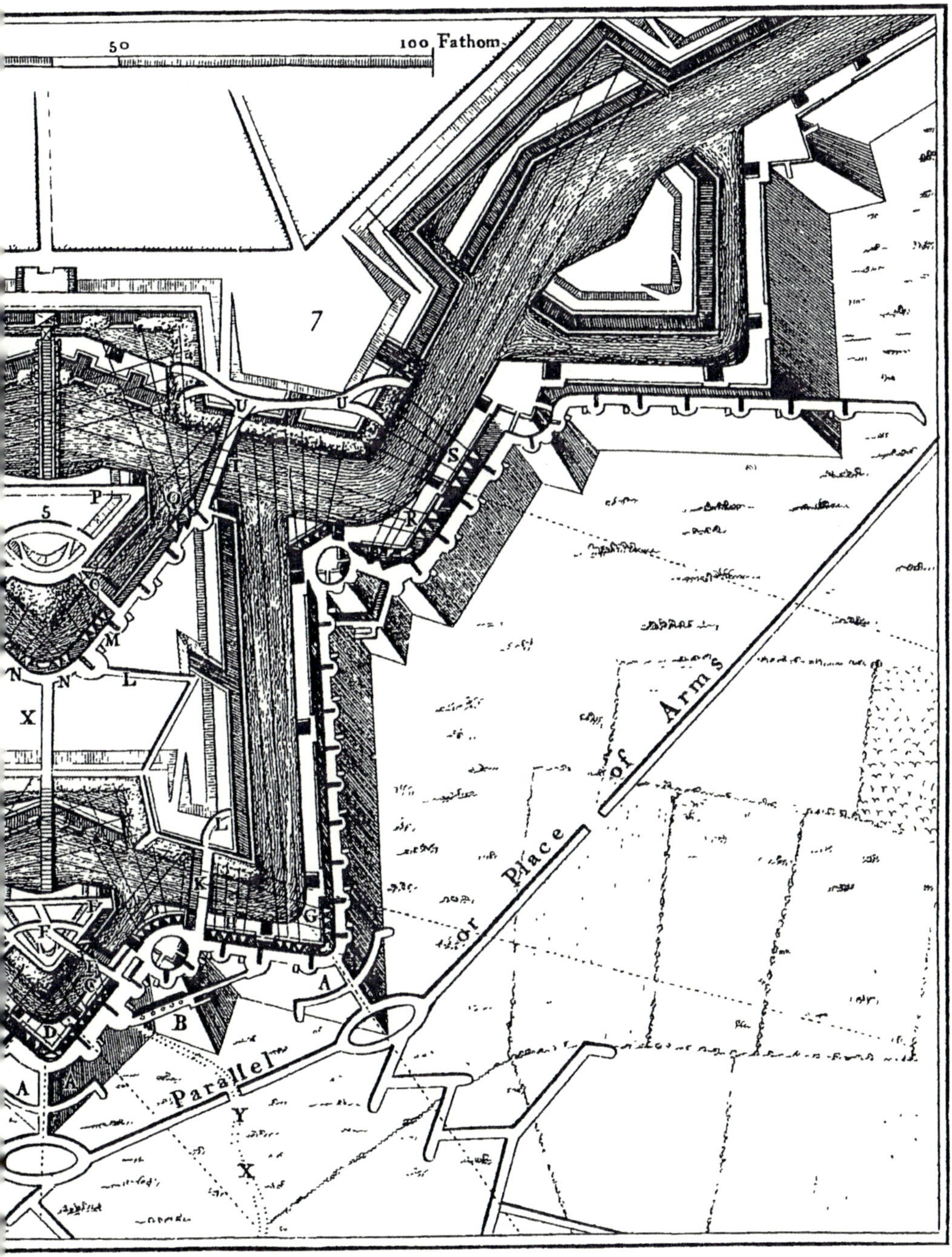
50
100 Fathom
7
U
U
S
T
P
5
O
R
Q
M
N
N
L
X
L
K
G
H
F
E
C
A
B
D
A
A
Parallel or Place of Arms
Y
X

which the garrifon will be fo ftraitened as not to be able to hold out long.

When a Town fituated on an eminence is furrounded with rocks, having little or no foil to cover them, the Attac becomes very troublefome, and next to impoffible; fuch is *Luxembourg*, where the only fide that can be attacked is all rock, covered with no more earth than a foot or eighteen inches; there is no other remedy to be found in that cafe, than to fupply the want of earth by wool-packs, gabions, fafcines, and fand-bags.

It often happens that thefe Places are cut into the rock, fo that the beft part of their revetements are of folid ftone; then the fixing the miners, or making Breach with guns, becomes very tedious and troublefome, and the more fo in proportion as the ftone is harder.

The rock muft be well examined to know whether there are no veins that are fofter than the reft, which is often the cafe; and when there are any fuch veins to be found, the work will go on much fafter: but on the contrary, if the rock is every where equally hard, there is no other remedy but time and patience. *Goulon* propofes making the defcent into the ditch, by a hole made near the counterfcarp fo deep as poffible; he fuppofes a ditch of 30 feet deep, and the hole about 6 or 7, which he fays may be done in 7 or 8 days; at the bottom of this hole two Mines are to be made, one at each fide; and before they are fprung, a great quantity of fafcines, ftones, earth, *&c.* are to be thrown into the ditch juft oppofite them, which, together with the rubbifh of the Mines, will make an eafy defcent.

To make Breach, he would have 7 or 8 pieces of cannon to batter the wall where the rock ends and

and the masonry begins, and from thence upwards, that the rubbish and earth together may make a slope of pretty easy ascent: if the Breach should not be sufficiently wide, the miners may enlarge it by working into the rubbish, and placing some Mines there, which, when sprung, will not fail of making the Breach the necessary width.

The ATTAC *of a Place near the Sea.*

THE Attac of such Places is not likely to succeed unless the Besiegers are masters at Sea, so as to block up the harbour, and prevent the Besieged from receiving any relief that way. When *Ostend* was formerly besieged by the *Spaniards*, it held out three years, the *Dutch* supplying it continually with necessaries of all kinds by Sea, because the *Spaniards* had no shipping to block up the harbour, whereby it was enabled to make that long defence; whereas the defence that the same Town made in 1745, was the least of that nature that ever was known, for it held out but 12 days from the day of investiture to the end of the Siege. The Besiegers found means to raise Batteries which commanded the entrance of the harbour; so that no ships or boats could either go out or come in without great danger. The bad condition of the sluices, neglect of the fortification, and a general want of every thing necessary in the defence, occasion'd the shortness of the defence.

It is not always sufficient for the Besiegers to be masters at Sea, and to block up the harbour, unless farther precautions are taken; for the Besieged may watch and take the opportunity of a dark or foggy night, to pass by the fleet unperceived, with small

ſmall boats, and ſupply the Town with neceſſaries, and therefore a boom, or ſeveral chains, ſhould be faſten'd croſs the entrance of the harbour; and for a farther ſecurity, ſmall boats with lighted torches ſhould row up and down along the ſhore all night; by which means it would be impoſſible for any thing to paſs by, to leave the Beſieged the leaſt hope of ſuccour that way. Sometimes a ſtockade is made croſs the harbour, which effectually ſecures the entrance; as the famous cardinal *de Richelieu* did at *Rochelle.* The Harbour being well ſecured, the Trenches are carried on as before directed; their figure and ſituation muſt be made as the nature of the ground will permit; in the mean time the ſhips firing both with their guns and mortars, will contribute much to a ſpeedy ſurrender. The Attac of ſuch a Place is generally made near the ſea ſide, where there are frequently hills of ſand blown there by the winds, under cover of which the Beſiegers may approach the Place very near, and make Batteries behind them, to annoy the works; this was the caſe at *Oſtend* in the laſt Attac, the Town being ſurrounded with water every where, except at the north ſide near the ſea, where there are ſand-hills reaching almoſt quite to the out-works, behind which the *French* erected Batteries ſo near the Place, that the Beſieged were ſo ſoon obliged to ſurrender.

To prevent Succours being thrown into a Town.

I Have hitherto ſuppoſed the General to have taken every neceſſary precaution to prevent the enemies throwing any ſuccour into the Place, and to be in a condition to withſtand any attempt that may

may be made to relieve it; but as I have not yet entered into any particulars, it will not be improper to expatiate a little on this uſeful part of the attacs, upon which the ſucceſs of a Siege very much depends.

It has happened ſometimes, that the enemy who had been thought in too weak a condition to relieve the Place, were afterwards found able to either raiſe the Siege, or to throw ſuch ſuccours into it, as to make it a hard task to take it: the enemy may be ſtrengthened either by drawing out moſt of the troops from the neighboring garriſons, which are leaſt expoſed, or by a ſtrong reinforcement received from ſome other part. In ſuch caſes,

There are two ways of preventing ſuccours being thrown into a Town beſieged, the one by remaining in the line, to hinder the enemy from breaking thro'; and the other, by drawing out as many troops as can be ſpared, without leaving the Trenches and Line unprovided againſt any ſally the Beſieged make, to engage the enemy upon an advantagious ſpot of ground.

Both theſe ways have been practiced, and both ſucceeded ſometimes, and at others miſcarried; ſo that it is hard to judge which of them is preferable to the other. The inconveniency ariſing from keeping within the Line is, the uncertainty on which ſide the enemy intends to make the Attac; whence ariſes a neceſſity of guarding equally all the poſts, which when the Line is pretty extenſive, or the Army of the Beſiegers not very ſtrong, gives the enemy an opportunity of attacking any part with great ſuperiority and advantage; and when once they have broken in, the Beſiegers placed there, inſtead

inſtead of charging them, yet in diſorder, uſually give all over for loſt, and taking to their heels, ſet an example for all the reſt to follow; whereby the enemy ſucceed to their utmoſt wiſhes, without any farther trouble or danger. This has been the caſe of moſt Lines that have been attacked and forced.

The inconveniences ariſing from going out of the Lines, to engage the enemy at a diſtance, are, that the army muſt be divided, ſo as that one part may guard the Trenches, whilſt the other meets and oppoſes the enemy; which weakens it ſo conſiderably, that if they are pretty ſtrong, they may be in a condition to beat the Beſiegers; or if they do not think proper to engage, may draw them one way whilſt a ſuccour paſſes another.

If the Beſiegers are determined to ſtay within the Line, the General ſhould take care that it be every where in a good condition; the bridges of communication every where in good order, and of a ſufficient number; that the troops may be in a condition to ſuccour one another, upon occaſion, without any impediment; out-works ſhould be made in thoſe avenues which are moſt expoſed and likelieſt to be attacked; that the Line be every where equally guarded, thoſe parts excepted only which are inacceſſible, as ſome generally are; to keep a ſtrict watch upon thoſe parts where there are any hollow ways or riſing grounds, under cover of which the enemy may approach unperceived; to place redouts even at ſome diſtance beyond the Line, if the ſituation ſhould be favourable to the enemy; parties and ſpies to range continually about, to give timely notice of the enemy's approach; for if the Line is well made, and the

enemy's

enemy's deſign diſcovered betimes, any attempt upon it may certainly be prevented.

In a country where wood is to be had, fires may be made in the night without the Line, about 100 yards diſtance, oppoſite to every ſaliant angle of all the redans, whereby the approach of the enemy will be diſcovered, and his diſign fruſtrated.

But if, notwithſtanding all the precautions taken on the part of the Beſiegers, the enemy ſhould reſolve to attempt the relief of the Town, by attacking the Line in ſome part or other, a ſignal muſt be made to notify their approach, and on which ſide; whereupon the troops next at hand immediately march to the aſſiſtance of thoſe attacked, to keep the enemy in play till others can come up, which are at a greater diſtance; then vigorouſly charging the enemy upon their firſt entering the Line, while they are yet in diſorder, and, if poſſible, to drive them out again. It will be neceſſary to keep guards, at the ſame time, all round the Line, at proper diſtances, leſt they ſhould make an attac on one ſide, whilſt they attempt to throw in ſuccours on another.

I am perſuaded that if the Beſiegers knew their advantage, and did not give all over for loſt, as ſoon as the enemies are broke in, and would charge them briſkly before they have time to put themſelves in order, few or no Lines would be forced, ſuppoſing all the precautions before-mentioned were obſerved.

But if the Line ſhould be too extenſive for the number of the Beſiegers to defend, or the enemy ſhould appear and be ready to attac them, before the Line and bridges of communication are finiſhed, in that caſe, it would be moſt convenient to march

out

out and attac the enemy before they advance too near, and before either thoſe in the Town have any notice of their arrival, or they themſelves find means of throwing ſome relief into the Town.

So ſoon as intelligence has been received of the enemies approach, a convenient ſpot of ground muſt be choſen for the field of battle, if poſſible either near ſome defiles, through which they muſt neceſſarily paſs, or ſome place where Batteries may be erected on a hill or riſing ground, or elſe near ſome village, where the troops being poſted, may flank the enemies in their Attacs. The knowledge and skill of the General is here to be diſplayed, to procure all the advantage which the nature of the ground can afford; and the diſpoſition of his Army ought to be ſuch, as not to be either ſurprized or flanked by the enemy, but, on the contrary, to afford mutual ſuccours to one another, whenever it is neceſſary.

REMARKS.

THERE are various opinions concerning the Defence of Lines, but the greater number is on the ſide of thoſe, who think it is more advantageous for the Beſiegers to meet their enemies, and fight them before they come too near, for ſuch and other reaſons as have been mentioned before; but to ſet the one and the other manner into as clear a light as can be done, we ſhall here give ſome examples, both of Lines which have been forced, and others which were not.

When Prince *Eugene* beſieged *Belgrade*, the *Turks* came with an Army of 200,000 men, and beſieged him in his camp, making Approaches in the

the ſame manner as if before a Town; yet though his army was only 40,000, and the *Turkiſh* garriſon nearly equal to his whole army, he marched out of the Line in a foggy night, about three in the morning, attacked the advanced guards of the *Turks,* drove them back upon the corps that ſhould have ſuſtained them, and theſe falling back upon the others, put their whole Army into confuſion, which afforded Prince *Eugene* an opportunity to rout them; a great number were ſlain, with their general; all done with ſo much expedition and ſecrecy, that he was returned to his camp before the garriſon knew any thing of the affair, which ſurrendered the day following.

When the *Turks* beſieged *Vienna* in 1683, and the chriſtians appeared under the King of *Poland* and the Duke of *Lorrain,* to relieve it, the *Turks,* truſting to the ſuperiority of their Army, kept within their Line, but were defeated, loſt all their Artillery, baggage, and the greateſt part of their Army.

The Lines the *Spaniards* made at *Arras* were the ſtrongeſt ever known, yet they were forced, by Meſs. *Turene* and *Laferté,* with very little loſs on the ſide of the Aſſailants, who had made three different real Attacs, beſides ſome falſe ones.

At the Siege of *Turin* in 1706, Prince *Eugene* forced the *French* Lines with 72,000 men in them, himſelf having 24,000 only; he attacked them in a poſt where the Lines were neglected, and the communication of that quarter with the others very bad.

It appears from experience, that it is ſeldom in the power of the Beſiegers to prevent the Town from being relieved, without an Army of obſervation;

vation; in the few inſtances following, where the Lines have not been forced, it was chiefly owing to the great ſuperiority of the defenders; as at *Philipsbourg* in 1734, the *French* had a numerous Army, the Line of a ſmall extent, and extremely well made; whereas Prince *Eugene* had no more than 30,000; ſo that it would have been a very great imprudence to have attempted any ſuch thing.

But when the Beſiegers have an Army of Obſervation, it is hardly poſſible to force them, and relieve the Town; the only thing that the Beſieged can do then, is to attac a Town of the Beſiegers, whereby they will be either obliged to raiſe the Siege, or ſee a Town of theirs taken.

When *Liſle* was beſieged in 1706, Prince *Eugene* commanded the Attacs, and the Duke of *Marlborough* the Army of Obſervation; whenever the *French* made any motion towards relieving the Town, the Army of obſervation was immediately reinforced by as many troops as could conveniently be ſpared from the Trenches, which prevented the *French* from venturing to attac the Allies, and tho' they had a numerous Army, were obliged to remain quiet ſpectators of the Town's being taken.

Of the raiſing a SIEGE.

NOTWITHSTANDING all the precautions that a diligent and knowing General may poſſibly have taken, before the Attac was made, to prevent any relief being thrown into a Town, it will ſometimes happen, that the Enemy, by ſome means which could not poſſibly have been foreſeen, may become ſo ſtrong, as not only to be in a condition to raiſe the Siege, which is oft

done

done with an inferior Army, but alſo to give battle with a ſuperior force; then it is prudent in a General to raiſe the Siege, and retire betimes as privately as he can, thus:

Firſt, to draw off the guns and mortars from their batteries; then the tools are put into the waggons at dusk, the Artillery ſets out firſt, then the baggage, the guard remaining in the Trenches, firing in the ſame manner as if the Attac continued; and when they are advanced pretty far on their march, the troops follow in a regular order, and the guard of the Trenches draws off by degrees, and follows after; the fires in the camp are all the while left burning as uſual, to conceal their retreat from the garriſon, who otherwiſe might fall on their rear.

When there is no danger of a purſuit, the Artillery, baggage and Army draw off by day, with colours flying and drum beating, in the order ſpoke of; but ſhould it happen that the enemy, by forced marches, come unexpected, ſo as to leave no time to carry off every thing, then what remains muſt either be burned or deſtroyed, that it may not be of uſe to the enemy.

The Attacs of FORTS, CASTLES *or* HOUSES.

IN the courſe of a war, it oft happens that places of little importance, yet neceſſary to be taken on account of their guarding ſome paſſes, or ſerving to haraſs the foragers, fall on convoys, or exact contributions, &c. As ſuch places do not require the attention of a whole Army, ſome officers with detachments are ſent to take them.

These

Theſe places, for the moſt part, are no otherwiſe fortified than by ſingle walls, without a rampart, ſometimes with a ditch about them, and a ravelin or redoubt paliſſaded, to cover the gate or entrance. A detachment ſent on ſuch undertakings, ſhould be provided with ſome ſmall field-pieces, and a few ſmall mortars, and ſet out ſo as to arrive before the place about dusk; they immediately make ſome epaulement near the place to be attacked, to cover their troops, guns and mortars, and ſo ſoon as they can ſee, begin to fire very briskly till a breach is made, or the enemy think proper to ſurrender.

If guns and mortars cannot conveniently be tranſported, on account of bad roads, or of the great expedition which is oft required, to ſurprize and leave no time to the Enemy to ſtrengthen the garriſon, or elſe there is an expectation of doing the buſineſs without them, it will be of importance to be well acquainted with the place, and the nature of the ground about it; and ſo ſoon as the detachment is arrived, to make a falſe Attac on one ſide, to draw the whole force and attention of the enemy that way, while a party ſcales the wall, or enters by any other means on the other; and if they ſucceed, march directly towards the gate, break or cut it open, and then fall ſuddenly upon the garriſon, and attac them vigorouſly till they ſurrender; but in caſe they ſhould not ſucceed, a retreat is beat to retire in the beſt manner poſſible.

Petards are oft of great uſe in theſe kind of Attacs, for breaking open the gates, forcing barricades or paliſſades; if they can by any means be fixed, they never fail of their effect.

Feuquier carried off many garriſons, both in *Germany*

many and in *Italy*, by them, which could hardly have been done by any other means.

If a Caſtle ſtands upon a high hill, ſo that the guns can have no effect upon it, or the ſhells thrown into it, oblige the enemy to ſurrender, as is very oft the caſe in *Germany* or *Spain*, where moſt Caſtles are built upon high places; then a large Mine muſt be made, if the Caſtle is of importance enough to be taken, which being loaded with a great quantity of powder, and before the Mine is fired, notice is given the garriſon of their danger, which generally makes them ſurrender.

In king *William*'s war the *French* made a Mine under the Caſtle of *Alicant* in *Spain*, and when it was loaded they ſummoned the garriſon, who were *Engliſh*, to ſurrender; but ſome of the officers, to ſhew their bravery, thinking themſelves ſecure, were ſwallowed up in the gap which the rock made, with their Punch-bowl.

If a ſmall Place is neither to be ſurprized or forced, by any of thoſe methods, ſome Infantry muſt be placed in ſuch a manner as to fire continually at the enemy, wherever they appear, and in the mean time the gate muſt be forced open, either with a Petard, or cut with hatchets; but this muſt be done ſuddenly, before the enemy have time to conſider of what is doing at the gate, to ſecure the grenadiers in this dangerous undertaking.

If the enemy ſhould have retrenched or barricaded themſelves in a Houſe, Farm, or Church, which is not to be forced, without ſome kind of ſmall Artillery, it may be ſet a fire to make them ſurrender. Unfair as ſuch ways of proceeding may appear, war is an excuſe, that allows of every ſtratagem.

tagem. The *Hungarian* huſſars have oft practiced this method with ſucceſs.

The ſurprizing large Places.

IF the Governors were continually upon their guard, and ſtrictly obſerved military diſcipline day and night, it would be a hard task to ſurprize a Place ſo guarded; but it oft happens that a Governor lives in too great ſecurity, and neglect of diſcipline, eſpecially when the enemy are at ſome diſtance, and other Places between them; of which an active and skilful General knows how to take advantage, and to lull the Governor in his ſecurity, by giving out, in a private manner, that he intends to undertake ſomething quite contrary, in the mean time he is preparing every thing neceſſary for his expedition; he ſhould be informed, by ſpies, of the ſtrength of the Place and garriſon, what troops, whether Veteran or militia, all the centry-poſts, the guard houſes, and ſtrength of each guard, where the ſoldiers are quartered, in baracs or private houſes, the ſituation of the Governor's houſe, and of the ſtore-houſes, the place of parade, where the troops aſſemble, whether any poſts are guarded with leſs diligence than others, on account of their diſtance from the parts where they imagine the enemy may be expected; it is alſo neceſſary to be acquainted with the courage and capacity of the Governor, and of the principal officers, the nature of the ground about the Place, whether there are any woods, hollow roads, under cover of which the troops may approach unperceived. This being known, and the troops before the Place, while yet dark, it muſt

be

be determined where and which way to enter the Town; or ſometimes in foggy mornings, when the gates are opened, by lying hid in troops at hand, approach inſtantly and ſeize on the draw-bridge, to prevent its being drawn up, and keep the guards in play, if they cannot ſurprize them, till others may arrive; or get a cart loaded with fruit, which they ſtop, under ſome pretence, by the guard-room, let them fall out to amuſe the guard, which they ſeize, being under diſguiſe of peaſants, or ſtop under the draw-bridge, and drop one of the wheels, to prevent its being drawn up; or ſometimes this is to be done by means of ſome old aqueduct or underground paſſages that has been neglected; ſometimes a number of officers and ſoldiers enter in diſguiſe, ſome one way and ſome another; when they are got ſafe into the town, they harbour either in public houſes or in thoſe of people, bribed to receive them, and ſo ſoon as it is dark, and the garriſon lies ſecure in their beds, all aſſemble at a place agreed on, and from thence march directly to the gates and other poſts, and ſurprize the guard, ſending word, or making ſignals to the troops without to approach and break open the gates; ſo ſoon as theſe are got in, they march directly to the barracs, to prevent the garriſon from taking to arms, and, at the ſame time, ſend a party to ſecure the Governor; others to take all the conſiderable poſts, ſuch as Store-houſes, powder-magazines, and eſpecially the ramparts, to prevent the garriſon from making uſe of their cannon.

Should there be any ſtrong hold within the town, as a Caſtle or Citadel, care muſt be taken that the garriſon do not throw themſelves in it; and in caſe of any reſiſtance from thence, the guns of the

town

town may be turned againſt it, to reduce it as ſoon as poſſible.

Cremona being garriſoned by the *French,* was ſurprized by Prince *Eugene,* who took the Governor and M. *Villeroy* priſoners; but a regiment of the garriſon happening to be accidentally under arms, by ſome miſtake of the time, made an alarm, and the officer of Prince *Eugene*'s party, who was to have given the ſignal to thoſe without, was killed in the firſt Rencounter, the troops without, not having the expected ſignal, and hearing the firing and noiſe in the Town, marched off giving all over for loſt; which obliged Prince *Eugene* to retire, after he had been nine hours in poſſeſſion of the Place.

As nothing is more inſtructive than examples of facts, I ſhall here give a few more.

Feuquier commanding at *Pignerol* in 1691, ſurprized *Savillan* with four companies of the King of *Sardinia*'s gens d'armes in it; in this manner; in the evening of a froſty night, he ſet out with 800 horſe, having 500 foot behind them, and made ſuch haſte as to arrive two hours before day; he approached with his foot to a detached baſtion, which he knew had no revetement, entered the ditch, paſſed over the ice, to the entrance of the Place, where he enter'd, having the lock broke open by a ſpy ſent before; then drew up his foot, and ſurprized the guard, broke open the gate to let in the horſe; which done, the garriſon was taken, without any reſiſtance, and carried to *Pignerol:* all executed in thirty hours, tho' theſe two Places were above fourteen leagues aſunder.

The ſame Officer being quartered at *Hailbron* in 1688, marched with 900 horſe and 800 foot to a little place called *Kreilſheim,* in the country of

Anſpach

Anſpach on the *Neker*, the garriſon conſiſted of two batallions, incloſed by a good wall, a ditch round it, and a Caſtle within; as it could not be ſurprized, he perſuaded the Governor to come out under ſome ſpecious pretence, upon which *Feuquier* commanded him to ſummon the garriſon to ſurrender priſoners of war, which they did.

This ſhews, how cautious a Governor ſhould be, and not leave his garriſon without being well eſcorted, or truſt an enemy without great neceſſity.

The ſame gentleman commanding at *Phortzheim* upon the *Lentz* in 1689, being much ſtraiten'd by two poſts of the *Imperialiſts*, the one above and the other below him; for they placed guards upon hills ſo near, that they could diſcover whatever came out of his garriſon: to get rid of ſuch troubleſome neighbors, at dusk he marched out with 600 foot, took a great circuit, and arrived before the gate of *Neubourg*, one of theſe poſts, which was leaſt guarded, about midnight; where paſſing over the bridge to the gate, told the centry, in *German*, that he came from a party, and deſired to let them in; whilſt they ſent for the key to open the gate, he amuſing the officer of the guard with ſtories, the Petard was fixed before it was perceived; they immediately fired at him and his men, and beat the alarm; but it was too late, for they got in, and were drawn up before the garriſon could be brought under arms, and ſo were taken, conſiſting of 500 foot, 150 dragoons, and about 300 horſes.

The very next night he marched with 600 foot and ſome horſe to *Entzwahing*, which was the other poſt, being only guarded by 150 foot, and

500 horſe, ſending his horſe over the River *Lentz*, to prevent thoſe of the garriſon from running away, and attact both gates at once, knowing they were but ſlightly guarded, ſome cutting the paliſſades placed in the form of a redan to cover the gates, whilſt the reſt fired ſmartly at the guards; an opening being made, the gates were forced and the garriſon taken.

Theſe two garriſons, conſiſting of 1300 men were put to the ſword, as a ſacrifice to the manes of 30 troopers which the *Germans* had cut to pieces in cold blood; a terrible revenge of a provoked and cruel enemy.

The King of *Sardina* having a company of gens d'armes in the Caſtles of *Orbaſſan*, a league from *Turin*, to cover his going to a pleaſure-houſe he had there; *Feuquier*, always upon the watch, finding they thought themſelves ſecure, and neglected their duty; ſet out from *Pignerol* in the dusk with 800 horſe and 500 foot; and when he came near the caſtle, detached his horſe towards *Turin* and *Montcallier* to watch the enemy who might come that way, and advanced with his foot to the gate, where he ordered a Petard to be fixed; the centry perceiving it, killed the Petardier and alarmed the garriſon; however, he himſelf fixed the Petard, as having no body elſe that underſtood the buſineſs; by which the gate being broke, he took the garriſon in the midſt of the King's quarters, and carried them ſafe to Pignerol, without the leaſt purſuit.

The

The taking a Place by Efcalade.

THE manner of taking a Place by Efcalade, is much the fame as that of furprifing it by any other ftratagem; the only difference is, in paffing the ditch and mounting the rampart by means of ladders. The fcaling ladders ufed, upon thefe occafions, are of various forts; fome are of ropes, and fome of wood, fome are made of feveral joints, fo as, when put together, to make a ladder of any length, which, in my opinion, are the beft, for the height of the wall is feldom known till you come upon the fpot; therefore no proper length can be given to the ladders before-hand; there is another fort ufed in *England*, much of the fame make as the common ladders, only the fteps turn about wooden pegs, fo that the poles may be brought near each other, or to fhut like a parallel ruler: this ladder is very convenient for carriage, but as they are of a certain length they are not fo ufeful as thofe with joints.

Being arrived before the place in the night, the firft thing to be confidered is, where and in what manner to pafs the ditch; when it is dry and deep, there needs no other confideration than how to get into it; if it is muddy, boards, hurdles, or fafcines are to be thrown in; but if it full of water, the paffage is like to be troublefome, but it often happens that a Governor, becaufe the Town feems to be in no immediate danger of a furprize, grows carelefs in his duty, and negligent of difcipline, and by that means may be eafily furprized.

When a river paffes by or through a Town, a great number of boats muft be provided in as private

a manner

a manner as poſſible, and brought in the dark, to be ready to carry the troops over, in the middle of the night, or early in the morning, about an hour before day.

But if there is a deep wet ditch, which has no communication with any river, ſmall boats made of Tin ſhould be provided, each to hold one man only; ſometimes baskets covered with skins or oil-cloth, have been uſed on ſuch occaſions; theſe kind of boats being very light, are eaſily carried by the detachment; and when the firſt have paſſed the ditch, they puſh the boats back for others to get over, till all are paſſed.

Suppoſing then the troops prepared to paſs the ditch, by ſome means or other, a party muſt firſt be placed on the counterſcarp oppoſite to the landing place, ready to fire at the garriſon, in caſe they have taken the alarm, and come to oppoſe their mounting the rampart. If the ditch is dry, the ladders are fixed in ſome place fartheſt diſtant from any centry, and ſo ſoon as they get upon the rampart, they put themſelves in order, to be ready to receive the enemy, if they ſhould appear; then the commanding officer, or ſome truſty man, who ſpeaks the language of the garriſon, advances ſome diſtance before the reſt, towards the gate; if he meets with a centry, he goes up to him, under ſome pretence or other, as if he belonged to the garriſon, and if the centry ſuffers himſelf to be thus ſurprized, claps a piſtol to his breaſt, to keep him quiet; but ſhould the centry, knowing his duty, offer to keep him at a diſtance, he muſt endeavour to kill him with all poſſible ſilence, and then advance ſuddenly with the detachment towards the gate, and either ſurprize or kill all who oppoſe them;

them; immediately upon this they fall to work breaking open the gate, to let in the reſt of the party, and then proceed in the manner before deſcribed.

If the ditch is wet, the rampart high, and has a revetement, it will be hard to ſurprize the Town that way; but if there is no revetement, the troops may hide themſelves along the outſide ſlope of the rampart, till all are over, and then proceed as before.

REMARKS.

IN Queen *Ann*'s war, *Keiſerſlautern* was ſurprized by the *French* in the following manner. A *German* deſerter told the *French* Commander, that if he would ſend a party with him, he would engage to ſurprize the Place; accordingly a detachment was ſent, which marched through the woods, till within half a league of the Place, where they ſtopped till it was dark, and provided themſelves with as many ſheep and horſe racks as they could get, to ſerve them for ladders. The ditch was dry, and the wall low; at dusk they approached the Town, got into the ditch, fixed their ladders; 400 grenadiers, provided with hatchets, mounted the rampart, with the *German* at their head; ſo ſoon as they were got up he advanced, at ſome diſtance before the reſt, to the centry, told him he went the round, that his light was out, and deſired him to ſtrike a light for him; the centry, not ſuſpecting any thing, went to ſtrike a light, and whilſt he was about it, the other gave him a blow with an iron bar, which he had ready for that purpoſe, by which he threw him into the ditch;

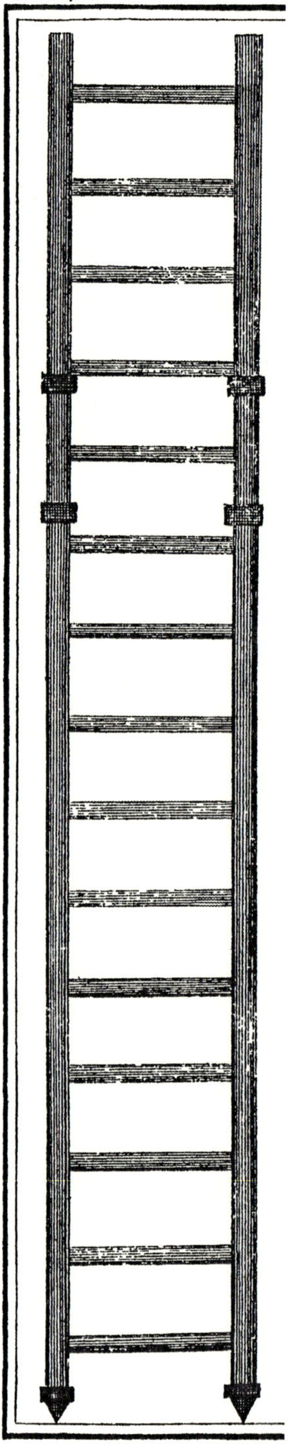

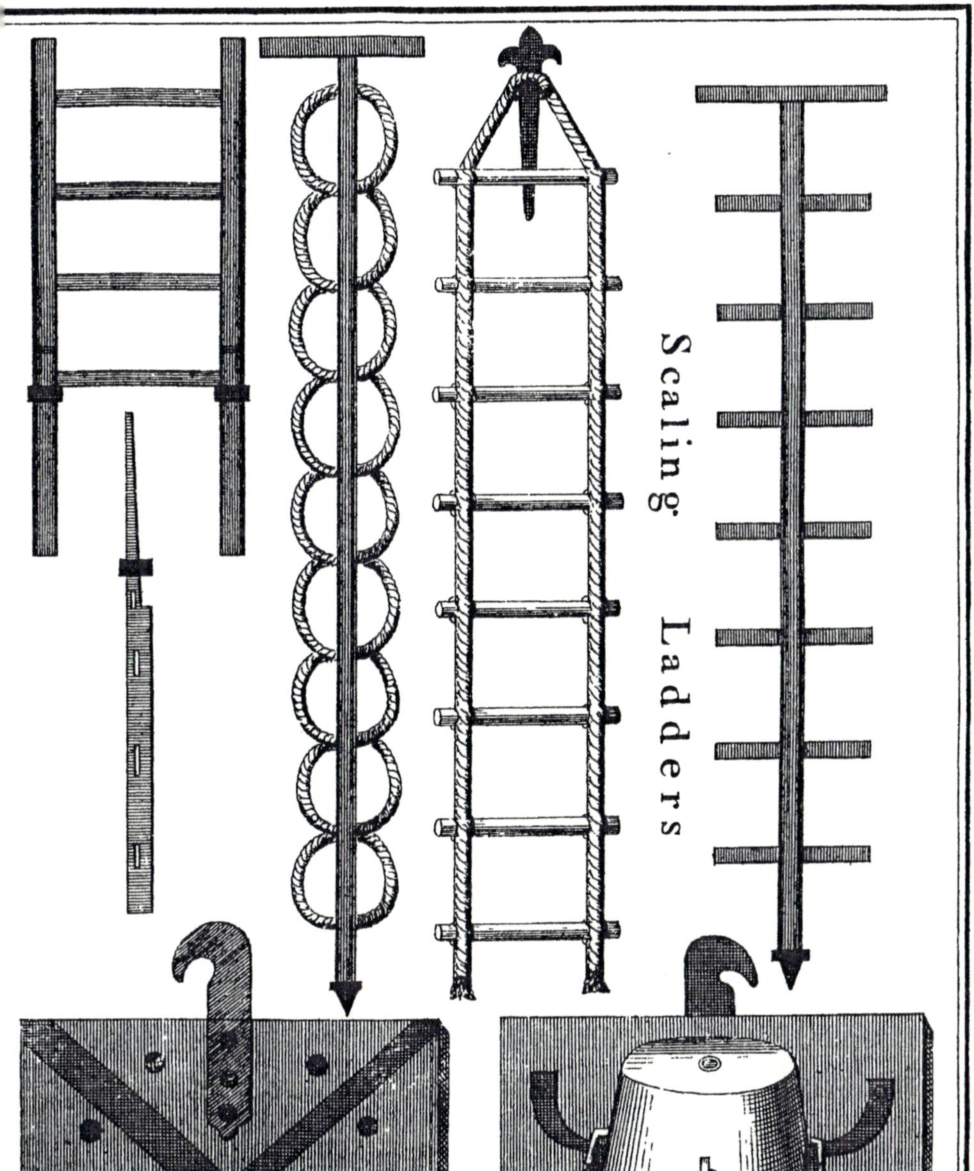
Scaling Ladders
A Petard

ditch; the grenadiers upon this immediately advanced to the gate, ſurprized the guard, and opened it. The garriſon finding how things went, withdrew into the Caſtle; and when the inhabitants had ranſomed the Town from plunder, the *French* retired, as they could not keep it for want of the Caſtle.

In the year 1676, *Loo,* a Town on the River *Dender,* belonging then to the *Spaniards,* was ſurprized by the *French.* The ditch was wet, and the rampart without a revetement; half the detachment paſſed the ditch in ſmall baskets covered with oiled cloths, while the other half ſtood ready to fire upon thoſe who ſhould oppoſe their paſſage, then followed the reſt in the ſame boats, and took the Place.

When the Army is very ſtrong, and the place very large, and not regularly fortified, ſeveral attacs may be made on different parts, at ſome diſtance from each other, ſome of them may be falſe ones, and begun firſt to favour the real ones; this may be done with ſcaling ladders, petards, and boats, if the whole ditch or part of it is dry. It was in this manner that the *French* and *Pruſſians* ſurprized *Prague* in the laſt war, juſt before a ſtrong reinforcement which was a coming, could get into the Town. Theſe and many other ſtratagems may be uſed to ſurprize a place, which ſeldom fails of ſucceſs, if the undertaking is carried on with reſolution, and well concerted. I am fully perſuaded, that many places that make a long defence when attact in a regular manner, might be taken at a ſudden attac on all ſides, if it can be before the garriſon is reinforced. It is true that ſuch an attac may be looked upon as deſperate, and it might expoſe the troops to great danger; but

but when we confider the great expence of a fiege, the time, and number of men loft, it will appear, that even by lofing the fame number of men, in a fudden affault on all fides at once, there would be much time and expence faved.

How and when a SIEGE *may be accelerated.*

WHEN a garrifon is very fmall, in comparifon to the extent of the Town, or the Army of the Befiegers very numerous, there is no neceffity for ftrictly obferving all the precautions mentioned in the Attac of a place regularly fortified; for which reafon the Line of circumvallation is oft omitted, as I am told the French did laft War at *Tournay*, tho' the Town was well fortified, and had a very ftrong garrifon; but the Befiegers were fo numerous, that they were under no apprehenfion of any attempt for its relief.

In fuch cafes an extraordinary number of workmen are employed, who carry the trenches within 100 or 150 yards of the Town, the firft night, and there the firft parallel is made, erecting likewife fome batteries with gabions and fand-bags.

When a garrifon is weak, the works of the Place out of repair, without any revetement, and but few or no out-works; or when there are any hollow ways or cavities near the Town; then, inftead of opening the Trenches at 800 fathoms diftance or more from the Place, as before, the Befiegers advance the firft night near or on the glacis, where they make a Parallel, and erect batteries; and when this work is fo forward as to cover the troops, then they work backwards; that is, they carry a trench or two from the Parallel already made back-

backwards towards their camp, for the troops going to or coming from the Parallel to the camp in ſafety, and at the ſame time carry their lodgments forward on the ridge of the glacis.

If the garriſon is very weak, the works made with turf, and out of repair, the Beſiegers may advance to the covert-way the firſt night, drive out the enemy, and lodge themſelves there, if it would prove too dangerous to aſſault the body of the place at once. It would be proper to carry ſome ſmall field-pieces along with them, to diſcommode the Beſieged, till their Batteries are raiſed, and a good number of wool-packs to cover their troops in the covert-way, till their lodgments are made ready to receive them. For it is not ſufficient to advance a ſiege only, without taking care, at the ſame time, to loſe ſo few men as poſſible, that the advantages gained one way, may not be counter-balanced by the loſs on the other.

When a Siege is undertaken, the firſt conſideration is, how to proceed ſo as to become maſter of the Place in the ſhorteſt time and ſafeſt way, ſometimes attacking it in a regular manner, when it cannot be effected otherwiſe, ſometimes advancing pretty near the Place, under cover of ſome riſing ground, hollow roads, caverns, or buildings, which the Beſieged may have neglected, or had not time to deſtroy; ſometimes by attacking the Town on one ſide, and ſcaling it on the other; or ſurprizing ſome of the out-works, eſpecially when they are made with ſods only. In ſhort, a ſkilful Commander ſhould uſe all kinds of ſtratagems to over-reach the enemy, and if not in the whole, at leaſt to ſurprize them in part, and never omit the leaſt favourable opportunity, ariſing either from

from the nature of the ground, indolence of the Governor, or ill condition of the work, which may any ways contribute to ſhorten the Siege. But as it is impoſſible to enumerate all the various caſes that may happen, and which depends chiefly on the time, opportunities, and ſituation, we ſhall inſiſt no longer on the Attac of Places, but proceed to their Defence.

PART

PART II.

Of the DEFENCE.

IN the firſt part I have explained the particulars of the Attac of Places, the ſecond treats of every thing concerning their Defence: there is ſuch a connection between theſe two branches, that it may not altogether appear improper to treat of them together in the ſame manner as *Goulon* has in his memoirs, were it not more proper with regard to the perſpicuity of the ſubject to ſeparate them; the Attac teaches the elements of Defence in general, I ſhall here treat more particularly of what may be uſeful in the Defence of each part, or in any manner conducive to a long and vigorous Defence of the whole.

The Deſign of this work being to point out the principal means to be uſed in the Defence of a Place; I ſhall begin with the neceſſary garriſon and quantity of ammunition and proviſion requiſite in proportion to its bigneſs, and the general diſpoſition which is to be made, to make a brave and ſtout Defence; after this we give the particulars relating to the Defence of each part of the Fortification, and then an explanation with regard to the Defence of ſmall Places, Caſtles, and even of Houſes, capable of receiving a guard, with a view to rouſe the attention of young officers, and to ſhew that there are few poſts but may be defended againſt an attac of a ſhort duration; and that theſe kind of Defences have their uſe, and are not alto-

toge-

together romantick, when well conducted by brave and skilful Commanders.

TROOPS *and* AMMUNITION *required in a* FORTIFICATION *to make a good* DEFENCE.

THE Defence of a Place depends not only on the goodneſs of its Fortification, but alſo on the ſtrength of the garriſon, the quantity of Artillery, ammunition, and proviſion; and above all, on the courage and conduct of the Governor.

It is not ſufficient that a Governor is brave, and ready to expoſe himſelf on all occaſions to the greateſt danger, but he ſhould have a perfect knowledge and skill in fortification, to make a ſtout and long Defence; by theſe he will be enabled to make the beſt uſe of every minute part of the works, and to defend them with leaſt danger to the garriſon, as likewiſe to make the enemy pay dear for every inch of ground they gain.

But how ſhould it be poſſible for a Governor to know after what manner to defend the ſeveral works with beſt advantage, if he is ignorant in what particulars their Defence conſiſts? to retire and give way to the ſuperiority of the Beſiegers upon occaſion, and at a more favourable conjuncture to return to drive them out again, and deſtroy their works? might not one conclude, without preſumption, that the little reſiſtance which ſome conſiderable Places have lately made, tho' well provided with every thing, and capable in all other reſpects to make a long and obſtinate Defence, has been owing chiefly to the want of knowledge and skill in the Governors?

It is certain that the ſucceſs or miſcarriage of all great undertakings, eſpecially in the Art of War, chiefly

chiefly depends on the knowledge and conduct of the Commander; ſince nothing is done without his orders, and every thing is carried on under his direction, both in the projecting and execution.

It is then to be taken for granted, that conduct and knowledge in the Governor is the firſt thing eſſentially neceſſary for a good defence, and next a ſufficient garriſon, well provided with every thing, both for defence, and enabling them to ſuſtain the great fatigues unavoidable in a Siege.

If it is of importance that the works ſhould be well conſtructed, it is no leſs ſo to have a ſufficient number of troops to defend them, ſince they are made on the ſuppoſition that there ſhould be men to defend them, and conſequently, the wall without Men, would be but little obſtruction to the enemy.

The bigneſs of a Place, the number of out-works, and the nature of the ground about it, ought to ſerve as a rule, whereby the ſtrength of the garriſon ſhould be determined. It is evident that the larger the Place is, that is, the more baſtions it has, the more troops are required for its defence; the ſame may be ſaid with regard to its out-works: and as to the nature of the ground about it, it is no leſs evident, that of two Places of equal bigneſs and equal number of out-works, that which is acceſſible on every ſide requires more troops to defend it, than that which is only ſo in ſome parts, ſince the inacceſſible ſides need not be ſo well guarded.

Mr. *Vauban* computes the ſtrength of a garriſon by allowing 5 or 600 men for every baſtion, regularly fortified with ravelins and a covert-way, and ſo many for every horn or crown-work, as likewiſe for a double ditch and covert-way; but this does

does not ſeem to be ſufficient for a Place acceſſible on every ſide, eſpecially now, when there is ſo much Artillery uſed in the Attacs of Places. There ſhould, in my opinion, be no leſs than 1000 men allowed for every baſtion, and in proportion for any other ſupernumerary outworks; and if there are any ſides inacceſſible, to allow only one third for each, that is, 333; beſides theſe, there is allowed one tenth cavalry; that ſo, according to this computation, a Place with ſix baſtions ſhould have a garriſon of ſix times 1000, or 6000 foot, and 600 horſe, or rather dragoons, who are beſt, as they charge equally afoot as on horſeback; that is, ſuch a Place ſhould have a garriſon of 6600 effective men.

It is likewiſe judged, that eight pieces of cannon and two mortars, are neceſſary for every baſtion; that is, 48 pieces of cannon and 12 mortars, for a Place of ſix baſtions, which in my opinion is not ſufficient; I would therefore allow 12 pieces for each baſtion, ſo that the whole will amount to 72 pieces of cannon and 40 mortars, in a Place acceſſible on every ſide.

As to gunners and bombardiers, there ſhould be four men to every cannon or mortar, at a time; and to have two days reſt, there muſt be 12 for every piece; that is 12 times 84 or 1008, or a battalion of 1000 Artillery men.

The number of Engineers muſt be conſidered. There ſhould be two to every batallion; ſo that 12 will be neceſſary.

Though this eſtimate is not ſo exact, as to be depended on in all caſes, yet it may ſerve as a pattern to go by, and may be increaſed or diminiſhed as occaſion requires.

As

As the Defences of Places equally fortified are ſo various with regard to the times, that there can hardly be any ſuppoſition made about them; and yet to make an eſtimate of the ammunition and proviſion, it is abſolutely neceſſary to be determined; for which reaſon, I ſhall ſuppoſe that the Place will hold out ſix weeks, as it may, if well defended, and make the following computation.

This ſuppoſition may not improperly be objected againſt, ſince the ſtrongeſt Places, that were attacked ſince the beginning of the war in 1740, *Fribourg* and *Bergen-op-zoom* only excepted, have not held out above a month; and hence it ſeemingly appears, that this time is more than neceſſary; but we are not to ſuppoſe, that a Place defended as it ſhould be, may hold out not ſo long; on the contrary, it may poſſibly hold out that time; beſides, it is always prudent to have plenty of every thing, and rather more than what is wanting, than not ſufficient.

The garriſon of a Place of ſix baſtions, being then ſuppoſed of 6000 foot, they ſhould be divided thus; the number of ſick and wounded in a fortnight's time, and thoſe reſerved for helping the gunners on the Batteries, and tranſporting the ammunition and other neceſſaries, may be computed at about 600; the remained 5400 ſerve for the Defence of the Place; theſe 5400 are divided in three bodies, of 1800 each, one of them for the guard, one to lie under arms, ready to march on the firſt notice where wanted, and one to reſt, but ready to march upon occaſion.

The horſe are alſo divided into three bodies, of 200 each; one for the guard, which are chiefly placed on the right and left of the Attac; the o-

thers

thers are divided into ſmall detachments, placed in different parts of the Town, to keep the inhabitants in awe, as well as to go upon any expedition; and the third reſt, with their horſes ſaddled, ready to mount for marching, if required.

The guard of 1800 foot may be divided into three bodies, of 600 each; two to defend the works attact, and the third to be placed in thoſe works which are not attact; the two firſt 1200 may be ſubdivided into three leſſer bodies, 400 each, two of which, 800, fire for two hours, the other relieves one of them at the end of that time, the next is relieved two hours after, and ſo on alternately, that after four hours firing they may have two hours reſt.

There is not ſo much firing required by day as by night; becauſe the Beſiegers cannot ſo well undertake any thing without the knowledge of the Beſieged; but in the night their proceedings are not ſo eaſily diſcovered; for which reaſon there muſt be a continual brisk fire all night.

By day the troops fire from between baskets, ſand-bags, or gabions, placed on the interior ſide of the parapet, and filled with earth, that being under cover, they may take better aim at their enemies.

To know the quantity of powder required for a Siege of any determined time, it will be neceſſary to know the quantity expended each day.

According to Mr. *Vauban*, a pound of powder of 16 ounces contains between 30 and 40 charges for common muskets, including their priming; each man may fire 100 charges during the guard of 24 hours; the quantity for each man may be fixed at two pounds at a medium; ſome will uſe more and ſome leſs.

I have

	lb.
I have ſuppoſed that out of the 1800 men for the guard, there were but two thirds placed in the front attacked, that is, 1200 men allowing two pounds for each man, amounts to — —	2400
For the two thirds of thoſe under arms ready to ſupport the guard, that is, 1200 men, at the rate of half a pound each	600
For the 600 men of the guard placed in the adjacent works of the Attac, at the rate of a quarter of a pound each	150
For the 200 horſe-guard, at a quarter of a pound each	50
For 300 cannon-ſhot during 24 hours, at 5 pounds each. —— ——	1500

The quantity of powder for cannon-ſhot may be determined exactly when the number of pieces and their callibers are known, by allowing a quarter or one third of the weight of their ſhot, for each charge.

For 300 ſhot for wall-pieces, at two ounces each. —— ——	38
Total lb.	4738

This is the quantity of powder expended each day, beſides what is uſed for mortars, grenades and mines.

Now, ſuppoſe the ſiege to hold ſix weeks, as ſaid before, the expence for 42 days, will amount to 198996 lb.

	lb.
Or the following quantity; a little more or leſs being of no great conſequence	200000
Beſides, for 1200 thirteen inch ſhells thrown during the ſiege, at the rate of 16 pounds each charge —— ——	19200

For

For 300 eight inch ſhells, and leſſer, at 6 pounds each —— ——	18000
For 6000 ſhots with ſtone mortars, at a pound and a half each ——	9000
For 3600 grenades, at 4 ounces each	900
For Mines —— ——	8000
Powder burnt in the breaches —	4000
For fire-works ——	4000
Allowance for waſte and accidents	24900
Quantity which ought to remain at the ſurrender —— ——	12000
Total lb.	300000

Having given an eſtimate of the quantity of powder required in a ſiege, the leaden ſhot is taken at two pounds for each pound of powder; and as there has been found 135996 pounds of powder for musket-ſhot, and for wall-pieces; twice that number, 271992, or 272000, will expreſs the number of pounds of lead.

As to the quantity of matches, there ſhould be 500,000 bundles.

Proviſion being of no leſs conſequence than ammunition, it will not be improper to ſhew how to eſtimate the quantity required of each fort.

Each man is allowed a pound and a half of bread a day, ſo that the garriſon conſiſting of 6600 men, they will require 9900 pounds a day, and 42 times that number, that is 415800, for the whole time of the Siege.

A common ſack of flower, containing four buſhels, weighs generally 200 pounds, and makes 270 pounds of bread; now the garriſon ſuppoſed to be 6600 men, at a pound and a half a day, amounts to 9900 pounds of bread required for one day, and therefore 415800 pounds for ſix weeks; which

which being divided by the number 270 pounds a ſack of flower makes, gives 1540 ſacks of flower required for 42 days ſubſiſtance of bread to the garriſon.

Tho' the garriſon diminiſhes daily, this quantity of bread is not ſufficient, becauſe each officer has ſervants, more or leſs, according to his rank; and there are beſides, women, ſurgeons, and ſeveral artificers of different kinds, that are uſeful in the Defence.

There muſt, beſides, be at leaſt, three days proviſions left when the garriſon capitulates, otherwiſe they are not entitled to a capitulation; for which reaſon, it will not be amiſs to have 2000 ſacks, in a Town beſieged, ſuch as we have ſuppoſed for a garriſon of 6600. The *French* allow the ſame quantity of bread to a garriſon of 3960 men; but then they ſuppoſe that each man has but half a pound of meat, which is too little, eſpecially for *Engliſh* ſoldiers, who can have no leſs than a pound and a half, amounting to 415800 pounds for the whole Siege.

There are likewiſe many other things neceſſary in a Siege, ſuch as roots, flower for pudding, herbs, pulſes, wine, beer, brandy, tobacco, &c. for the ſubſiſtance of the garriſon; and in general, every thing neceſſary for the Defence ſhould be provided with great care and judgment, that the garriſon may make all poſſible Defence.

Neceſſary Preparations before a Town is inveſted.

A Diligent Governor ſhould, in time of peace, conſider the condition of the Place, in ſuch a manner, as if he were going to be Beſieged, not omitting

omitting the leaſt thing that may contribute to a long and obſtinate Defence; he ſhould therefore take particular care to keep the fortification in good repair; clearing the country round of all hedges, ditches, trees, hollow roads, caverns, and riſing grounds, within the reach of cannon-ſhot, not ſuffering any houſes to be built within that diſtance, nor in general any thing to be done that may any ways favour the approach of an enemy.

He ſhould conſider well with himſelf, of every minute circumſtance that may be an advantage or diſadvantage to him during the Siege; he ſhould thoroughly examine the ſeveral works, and canvaſs all the different ſtratagems that may be uſed, either to defend them, or to give way upon occaſion, when overpowered, with intent to return and diſlodge the Enemy, after they have got poſſeſſion of them. In ſhort, how to defend his Place, inch by inch, with the beſt advantage, he ſhould conſider how, and in what manner the works defend each other, whether their communications are ſafe, or if they may be interrupted by the Beſiegers; how to incommode the enemies when they are at a diſtance, or to diſlodge them when near; and if the ground be proper for Mines, and where they ſhould be made; if any part of the country may not be laid under water, by means of dykes or ſluices; if there are any already made, to keep them in conſtant repair, or to make them, if they are wanted, taking care to place them ſo, that the enemy may not have it in their power to deſtroy them, either with their cannon or mortars.

If he is not ſufficiently skilled in the Attac and Defence, he ſhould frequently converſe with Engineers and others who underſtand it; entertain them

them at his table, and by engaging them to ſpeak freely, examine the works together, to ſee what may be done, to render the Defence of the Place as long as the circumſtances and nature of the works will admit of; and to make it familiar to himſelf, he ſhould ſet down a project of Defence on paper, and have it canvaſſed by the moſt skilful Officers and Engineers about him. This ſhould be done in private, that ſpies or deſerters, may not diſcover the weak parts to the enemies.

All theſe things being well conſidered, when the Town is in any danger of being beſieged, the Governor, beſides the attention of furniſhing the Place with ammunition and proviſion, ſhould likewiſe repair thoſe works that want it, and conſtruct, if poſſible, detached redouts, arrows, and ſuch other kind of out-works, which oblige the enemy to open their Trenches at a greater diſtance.

The covert-way ſhould be well paliſſaded; the ditches cleared of the mud, if they are wet; if they are dry, a trench ſhould be carried along the middle of them, ten or twelve feet wide, and five or ſix deep, and water brought into it, if poſſible; if any of the out-works have no revetement, but have paliſſades placed horizontally on the outward ſlope of their parapets, they ſhould be repaired and kept in good order.

The places of Arms in the covert-way may be retrenched, by raiſing parapets within parallel to the faces, and making a ſmall ditch before them. In ſhort, the Governor ſhould contrive all poſſible expedients to lengthen his Defence, and to oppoſe obſtacles to the enemy.

The galleries of the Mines are not to be neglected, they ſhould be made betimes, carried as

far

far into the field as the nature of the ground will permit, and their chambers placed under the ridge of the glacis, before the ſaliant angles of the covert-way, as likewiſe in all thoſe places where the enemy erect their Batteries.

If there are any houſes within the rampart, which may obſtruct the Defence, they ſhould be demoliſhed, and nothing left either within or without, which may any way favour the Beſiegers.

If there are any militia or new-raiſed troops in the garriſon, all poſſible care ſhould be taken to diſcipline them as well as time and opportunity will permit.

The Governor ſhould ſee that the Hoſpitals are in good order, and provided with every thing wholeſome and neceſſary, that the ſick and wounded may be taken care of, when there are any, nor think it beneath his character to viſit them, and to ſhew how much he is affected with their misfortunes; for ſuch inſtances of kindneſs, will not fail to engage every man, in time of danger, to do their utmoſt with pleaſure, to ſhew their gratitude, and to eſteem the honour, which the Governor gains by a good Defence, as their own.

As it is uſual in Sieges to throw a great number of ſhells into the Town, it is neceſſary to have ſome places under-ground bomb-proof, where the part of the garriſon not on duty may reſt in ſafety; they are not ſo much wanted in large Towns, where there are always ſome places or other ſecure from the enemy's ſhells, which is not the caſe in ſmaller ones, which therefore abſolutely require ſome ſuch ſecure places to be made.

They are generally made in the gorges of the baſtions, and ſometimes under the rampart behind the

the curtains. As ſoon as the enemy appears, they ſet up ſheds againſt the rampart, made with ſtrong timber roofs, covered with thick planks, and earth over them; ſome houſes fartheſt from the Attac are ſhored with ſtrong timbers, between every floor, from top to bottom, and the floors covered with ſtraw, wool, dung, faſcines, earth, *&c.* and theſe, when they are well ſecured, may either ſerve for the troops to reſt, or for hoſpitals.

The Powder magazines, above all, require a more ſpecial care: tho' they are built bomb-proof, yet when a great number of ſhells fall upon them, they ſeldom reſiſt their ſhock; for which reaſons, they ſhould be covered with ſeven or eight feet thick of earth, and a layer of faſcines, dung, and ſtrong planks laid over them; but if it ſhould happen, either from their ſituation or height, that this cannot be done, then a range of ſtrong timbers well faſtened together muſt be laid over the roof, to diminiſh the ſhock of the ſhells; their windows ſhould have no proſpect towards the enemy, and no body be permitted to go in or out of the doors, but at ſuch times as the fire of the enemy is ſlackened.

When there are no powder magazines in a Town, or ſuch as have not been made uſe of, it is a very troubleſome and difficult task to preſerve the powder from accidents; all that can be done in ſuch caſes is, to divide it into ſeveral parcels, and to place them in cellars and holes made under the rampart, at ſome diſtance from each other, and to cover them with planks and faſcines, having dung or earth laid over them.

The ſhells not only demoliſh the buildings upon which they fall, but likewiſe generally ſet fire to them;

them; to prevent this, a great number of casks ſhould be placed in the ſtreets, and the ſoldiers and inhabitants, being divided into ſeveral companies, a particular quarter allotted to each of them for extinguiſhing any fire which may happen there; by this means all confuſion will be prevented, and every part of the Town readily ſecured, in caſe of an accident. The pavements in all the ſtreets muſt likewiſe be taken up, and dung laid there, to prevent, as far as may be, farther damage and diſorder from the burſting of the ſhells.

The DEFENCE *from the Inveſtiture to the Attac of the* COVERT-WAY.

THE Place being inveſted, and the Beſiegers having begun the Line of circumvallation, the Governor ſhould fire at firſt, either with his ſmall pieces only, or with the great ones half loaded, to deceive the enemy, who imagining themſelves out of the reach of cannon, will advance and pitch their camp nearer the Town, which they ſhall have no ſooner done, but the garriſon is to give them a full diſcharge from their great guns, to oblige them to decamp once more, and remove to a farther diſtance.

If the country about the Place is divided by rivers or marſhes, as it generally happens, ſome quarters of the camp will be ſeparated from the reſt; in which caſe a Governor ſhould endeavour to fall upon ſome part with great precaution, for fear of the party's being cut off by the enemy's horſe; for the loſs of ſuch a part of his garriſon, at the beginning of the Siege, might be of bad conſequence.

As it is of importance to the Beſieged, to know on

on which ſide of the Place the Beſiegers intend to make their Attacs, no means ſhould be neglected to diſcover it. The magazine of materials, if it can be found out, may perhaps direct them in this particular; for it is to be preſumed, that the enemy will place them as near the opening of the Trenches as they poſſibly can; and therefore the Beſieged ſhould watch every night on that ſide, for the time of opening the Trenches, to be ready for firing briskly upon the workmen, while they are yet uncovered, which will very much retard the work.

So ſoon as it is known which ſide the enemy intend to attac, retrenchments ſhould immediately be made in the baſtions and ravelin that are to be attact, and not wait till the enemy comes too near, for then the fire of the cannon and mortars becomes ſo brisk, that it will hardly be poſſible to carry on theſe works; beſides, the Beſieged will then have other works on their hands, ſo as not to think on theſe.

During the time that the line is conſtructing, the enemy's engineers will omit no opportunity to get an exact knowledge of the Place, and the ground round about it, to form their plan of the Attacs, and which they will undoubtedly make upon the weakeſt ſide, if they can come at the knowledge of it. To prevent this, the Governor ſhould ſend 2 or 300 men every night, as ſoon as the Town is inveſted, to that ſide which he knows to be weakeſt, with orders to lie upon their faces, in the form of a ſemi-circle, whoſe diameter is, as it were, the paliſſades of the covert-way; theſe men ſhould be divided into ſmall bodies, of four or ſix men each, and diſtributed at about twenty

or

or thirty paces diſtance from each other, ſo as to occupy a large ſpot of ground.

A ſignal being agreed on, to give notice when any body paſſes between them, they ſhould remain in this poſture, and keep ſilence till day; if any body happens to paſs by, whoever ſees or hears them riſes and gives the ſignal to the reſt, who are to do the ſame; then all advancing to the covert-way, thoſe who paſſed are taken as it were in a net, without any poſſibility of relief from their eſcorte, who will hardly attempt to reſcue them from the hands of 2 or 300 men protected by the fire from the covert-way.

If thoſe who are thus reconnoitring, inſtead of paſſing through the intervals, ſhould fall in with ſome of theſe parties, and endeavour to get off, they muſt be fired upon; in which caſe, it will be difficult for them to eſcape either being killed or taken priſoners. If this precaution be rightly taken, it will not be an eaſy affair for the Beſiegers to get a true knowledge of the Fortification by reconnoitring.

Another precaution the Governor ſhould take, is to ſend every night, from the time the Place is inveſted, ſeveral ſmall parties, of ten or twelve men, commanded by a ſerjeant, with orders to lay on their faces, all round the skirt of the glacis, to keep ſtrict ſilence, and liſten with great attention to whatever paſſes.

What precaution ſoever the enemy may think proper to uſe, to conceal from the Beſieged the time and place of opening of the Trenches, it is hardly poſſible but the great motion, which they will be obliged to make about that time, muſt either be heard or perceived by thoſe placed in the covert-way,

way, who may even venture to advance ſomewhat nearer the enemy, obſerving ſilence, and great caution not to be ſurprized by the parties which the enemy ſend out on that ſide to obſerve that no one approach too near to diſturb their workmen.

The place of the Attacs being known, the great pieces of ordnance are mounted on the rampart of the Town to fire *en Barbet*, and the ſmall ones on the covert-way and other outworks, from whence they are to fire with all poſſible diligence; and to aſſiſt their aim, fire-balls are thrown from the mortars into the Trenches, to diſcover the workmen and their guards; in the moment of opening the Trenches, the greateſt fire ſhould be made upon the enemy as yet uncovered, and conſequently moſt expoſed.

The guns may continue to fire *en Barbet*, till ſuch time as the Beſiegers have erected their Batteries, which is commonly after the ſecond or third night of the Trenches being opened, but it will hardly be poſſible to keep them in that poſition long; afterwards, the Beſiegers fire being grown ſo much ſuperior, that the guns would ſoon be diſmounted, and therefore they muſt from thenceforwards be pointed through embraſures. The mortars and howitzes ſhould not be idle all the while the enemy are erecting their Batteries, but ſhould throw ſhells into the Trenches, and towards thoſe places where the enemy are moſt buſy; in ſhort, the beſt uſe is to be made of the Artillery before the Enemy are in a condition to ſilence them.

When the Enemy make a falſe Attac, it muſt be endeavoured to find out the real one, and to uſe the utmoſt diligence in making good retrenchments

in

in the outworks and in the baſtions; but as thoſe new-made works are eaſily deſtroyed, it had been much better to have made them in time of peace, that the earth may have time to ſettle; by which means they will be much ſtronger and firmer.

A Governor who underſtands Fortification as he ought, may eaſily judge which part of his place is weakeſt, therefore moſt liable to be attact, and conſequently he ſhould provide for its Defence in the beſt manner, by retrenchments and outworks, ſuch as he ſhall think proper.

The Artillery ſhould be diſpoſed in ſuch a manner as to enfilade the Approaches if poſſible, and fire continually upon the workmen at the head of the Trenches to retard the works. But when the batteries of the Beſiegers are once in a condition to return the compliment, it will be difficult for the Beſieged to maintain theirs, or prevent them from being deſtroyed; the ricochet batteries will enfilade the faces of the work oppoſite to the Attac, from one end to the other; neither will the traverſes which are made from diſtance to diſtance, be any effectual ſecurity againſt them, the ſhot falling in one place will bound over the next traverſe into another.

It would by no means be proper to fire with cannon at the Beſiegers batteries, who being ſuperior in number would ſoon deſtroy thoſe of the Beſieged, and conſequently render the guns uſeleſs for the remainder of the Siege; inſtead of which, ſo ſoon as the enemy have made a battery to diſmount the guns of the Beſieged, they ſhould be removed to ſome other place more ſecure, which may probably lead the Beſiegers into a belief that they have diſmounted them; and ſo ſoon as they

they have made another battery to diſmount them a ſecond time, they muſt be brought back to the firſt place or ſome other, where they may be out of danger of being diſmounted. This way of ſhifting the guns from place to place, whenever they are in danger of being diſmounted, will give the enemy much trouble, and make them loſe much time, which is all that is wanted or can be expected.

As mortars may be placed in any work without the enemy's knowledge, they may be uſed to very good purpoſe to deſtroy their batteries, eſpecially thoſe for battering in breach, or to diſmount the guns in the flanks; for the diſtance being ſo very ſhort, the bombardier muſt be very unskilful not to throw almoſt every ſhell upon them.

The beſt method of firing for the Batteries of the Beſieged is not in ſalvo's, or all at a time, for the Beſiegers place centries in the Trenches to look out and obſerve which way the guns are pointed, and give notice accordingly to thoſe who are in danger, that they may cover themſelves whilſt the reſt continue to work; but as this can only be done when there is an interval between the firings, whereas if the guns continue firing, though by ſingle pieces only, yet ſo as to leave no intermiſſion, and ſhells are thrown at the ſame time; the Beſiegers centries will not ſo readily venture to look out, and make their Obſervations which way the guns are pointed, by which means the workmen will work in fear.

Of SALLIES.

SALLIES, when made with due precaution and judgment, are one of the principal means to lengthen a Siege; whereas, when they are made without

without ſufficient care, or on unſeaſonable occaſions, they rather advance than retard the taking the Place; for which reaſon the utmoſt precaution ſhould be uſed in the execution, as well as great judgment in the choice of time and opportunities.

Sallies are proper when a Garriſon is numerous, or the Army of the Beſiegers is weak; for in ſuch caſes the keeping off the enemy from the Place as long as poſſible, may either diſpirit them, or retard the Siege to ſuch a degree, that they may be obliged to raiſe it, if the ſeaſon is far advanced, and the weather grows bad.

Sallies are alſo proper to be made when the Fortification is bad, or the garriſon ill-provided with ammunition and proviſion; it is beſt, in thoſe caſes, to keep the enemy at a diſtance as long as poſſible, to make the better terms with them, leſt coming too near the Place, and diſcovering either the wants of the garriſon, or the defect of the Fortification, they may have a chance to be made priſoners of war.

On the contrary, when a Place is well fortified and provided with every thing, Sallies are not ſo proper to be made, as the Garriſon may defend themſelves with greater advantage behind the works than in the open field; and what precaution ſoever may be taken to ſucceſs, yet ſome men muſt be loſt, and the Governor may be deprived of a ſufficient number to ſuſtain an aſſault, or to defend the works ſo well as they ought to be.

Sallies are dangerous when the enemy is at too great a diſtance, they may cut them off from the Town with their horſe; but when their Approaches are advanced between the ſecond and third Parallels, then is the time for the garriſon to ſally without

out much danger, the covert-way being at hand, from whence they are protected; they may even sally whilst the second Parallel is making, though with great caution not to be intercepted by the enemy.

But the most favourable opportunity for sallying is after the third Parallel is begun, and before it is finished; at that time the enemy can have but a small number of men at hand to sustain the workmen, who may easily be driven back, and their work destroy'd.

It is to be observed in general, that the most favourable time for sallying is when the works, which are to contain the guard of the Trenches, are imperfect, that is, before the second or third Parallels are finished, or the Places of arms between these Parallels are in a condition of receiving a sufficient number of troops to protect the workmen at the head of the Trenches. There are few actions in the whole Art of war, that require greater diligence, courage, and conduct than this; by diligence and courage the enemy are surprized and put in disorder or confusion, and oft obliged to abandon their works, which when once destroyed they can hardly ever regain, and re-establish; and by good conduct the enemy's own works may be turned against them; and when all the mischief that can be done is effected, a good and safe retreat is made; in short, good conduct secures from all the dangers which attends a contrary one.

Sallies may be made, either with a small or great number of men, that is, with ten or twelve men only, or with six, eight, or ten thousand; the intent of great Sallies should be to destroy a considerable part of the enemy's works, and thereby oblige

oblige them to repeat their labour to re-eſtabliſh them, to nail up their guns; to retake ſome poſt which had been loſt; and laſtly, to obſtruct and retard the enemy's works ſo much as poſſible.

As to ſmall Sallies, they ſerve to interrupt frequently the workmen at the head of the Trenches, who are ever ready to run away upon the leaſt alarm, deſiring no better pretence to leave their work; and as this is uſually done in the dark, moſt of them will get away if they can, and the reſt require ſome time to be brought back: if the enemy, after ſeveral alarms grow ſecure, and take no notice of them, they ſhould be followed by a more conſiderable one, well ſupported, which coming unexpected, will not fail having a good effect.

The beſt time for large Sallies is about two in the morning, as the troops being then much fatigued and ſleepy, are leſs capable of making any conſiderable reſiſtance, eſpecially againſt a vigorous and well-conducted Sally; and if it ſhould happen to rain ſo hard, as the guard of the Trenches ſhould not be able to make uſe of their fire-arms, the conjuncture would be too favourable to be neglected; in ſhort, no opportunity ſhould eſcape a ſagacious Governor to ſurprize the enemy, Sallies ſeldom or ever ſucceed in any other caſe.

Mr. *Vauban* would have the following Order obſerved in large Sallies; there ſhould be a detachment of 90 men drawn up, ſo as to have 30 men in front and three deep; to theſe are added a fourth rank of 30 grenadiers; the three firſt ranks ſhould be armed with breaſt-plates, and for weapons, a ſword and piſtol at their ſides, and a partizan or long iron-fork with a hook in their hands; this detachment ſhould be followed by another of

180 men, thirty in front and ſix deep; the firſt rank of which to be armed as the former, and the reſt as uſual; after theſe 200 workmen are to follow with the proper tools to demoliſh the enemy's works, ſome with combuſtibles and fire to burn what can no otherways be deſtroy'd, and ſome with long nails and hammers to nail up the cannon if an opportunity offers.

There ſhould be another detachment of 3 or 400 men ordered to ſupport the firſt in caſe of any attac from the Beſiegers, and to ſecure their retreat; theſe laſt are to follow the former in a ſlow pace as far as the head of the Trenches, where they halt, and ſtand ready to aſſiſt the others upon the firſt occaſion.

So ſoon as the guard of the Trenches are overthrown, the workmen fall to demoliſhing the works as faſt as poſſible, and if theſe troops can penetrate ſo far as the enemy's batteries, they nail up their cannon: when they have done all the miſchief that the time will permit, they retreat in good order; and if the enemy ſhould purſue them to the works of the place; then ſo ſoon as the Sally is got within the covert-way, all the cannon from the place ſhould fire briskly upon them, whilſt the troops in the covert-way and adjacent out-works, pour in their ſmall ſhot amongſt them, this will ſoon oblige them to retire, with conſiderable loſs.

In all actions performed by night, the troops ſhould have a mark to know one another, a white paper or handkerchief in their hats.

The troops deſigned for a Sally are drawn up in the covert-way, or in the ditch if it is dry, or elſe in ſome of the out-works, from whence they may be able to march out in great order.

If

If there is more than one attac, as there generally are, then Sallies may be made upon each of them at the ſame time, that the enemy finding themſelves ſuddenly attacked in ſeveral places at once, added to the terror heightened by the darkneſs of night, may throw their troops into confuſion, which the aſſailants obſerving, are to make their beſt advantage of: It would not be improper if ſome ſmall parties were ſent out ſeveral different ways, making a great noiſe to draw all the attention of the troops toward them, and then the great Sallies to fall on ſuddenly ſome other way, which ſeldom fails of having a good effect.

When the Beſiegers works are yet far off, two detachments of horſe are ordered, beſides the foot, on the right and left, to ſupport the Sallies, and prevent the enemy's horſe from cutting off their retreat; but after the third parallel is finiſhed, the ſallies are then made with foot only, and ſhould be ſtrong and frequent, provided the garriſon is numerous enough to diſpute every part of the ground.

So ſoon as the troops are retired into the covert-way, fire-balls ſhould be thrown into the Approaches, to diſcover the workmen, as they are then buſy in repairing the works, which were deſtroyed, and are, for the moſt part, uncovered, the fire of the place, well ſerved, at that time, may do them great miſchief, and retard the Siege conſiderably.

The ſmall Sallies, whoſe object are to diſturb the workmen frequently, without being in a condition to do them much hurt, are made by parties of ten, fifteen, or twenty ſtout men only, as ſaid before, who advance in great ſilence to the head of the Trenches,

Trenches, jump ſuddenly into them, making a great noiſe, and throwing ſome hand grenades, this done they retire, making the beſt of their way, for fear their ſmall number ſhould be diſcovered.

Of the LINE *of* COUNTER-APPROACH.

THE Line of *Counter-approach* is a kind of Trench, of 12 or 15 feet wide, which the Beſieged make from the covert-way, ſuch as a, a, at the right and left of the attacs, to enfilade the Trenches of the Beſiegers.

PL. XXI.

This Line has been neglected of late, not ſo much on account of its lengthening the Siege, as from the indolence and inactivity of the Governors, who think it ſufficient to ſecure their reputation if they keep a brisk firing with their great and ſmall arms, and make a few ſallies; and ſo ſoon as the covert-way is taken capitulate, to have the honours of war allowed them.

As this work is intended for thoſe brave officers who are deſirous to uſe their utmoſt endeavours to defend the Place, for their ſake I ſhall explain the uſe of theſe kind of Lines.

They are to be made in ſuch a manner, as to be enfiladed by ſome of the works belonging to the Town, and produced ſo far as to enfilade ſome part of the Trenches.

They may begin either at the ſaliant angles of the covert-way, before the adjacent ravelins to the Attac, or from the place of arms, as repreſented here, next to the ravelins; ſmall guns ſhould be placed next to them, to defend their approach by the enemy, and large ones upon the works that ſee in them to enfilade them, in caſe the enemy ſhould attempt

attempt to make a lodgment, after they have got poſſeſſion.

But it would be much better to carry a gallery underneath, to blow them up ſo ſoon as they are no longer ſerviceable. The enemy will endeavor to cover their works from being enfiladed by theſe Lines, by making frequent turnings, which cannot be done without loſs of time, and will carry Trenches ſo as to get poſſeſſion of them; but a few ſmall guns placed there will make their Approach dangerous. Another advantage of theſe Lines is, that they render the enemy's horſe uſeleſs againſt the Sallies of the Town, and even when they have covered themſelves from the enfilades of them, others may be made farther off, to enfilade them again.

Light field-pieces may be placed in theſe Lines, to enfilade their Trenches, which when near, grape-ſhot muſt be uſed, by which the Beſiegers will be obliged either to attac the Lines ſword in hand, or to make their Trenches deeper than ordinary; in either caſe the approaches will be retarded; but when the Beſiegers come once ſo near as that they may ſtorm theſe Lines, the pieces muſt be withdrawn for fear of being taken.

It may be obſerved that all work in general, which oblige the Beſiegers to open their Trenches at a diſtance or enfilade them, are very uſeful in retarding the Siege, which is all that is wanted or can be expected.

DEFENCE *of the* COUNTERSCARP.

THE enemy being arrived at the foot of the glacis, they muſt be diſturbed ſo oft as poſſible in their works, by firing at them from the covert-

covert-way, and from all the neareſt adjacent works; by ſallies, of a good number of men, to make them quit their works, to deſtroy them; and many fire-balls ſhould be thrown all night by which they will not be able to ſtir, without being ſeen and expoſed to the fire of the Place.

If there are any works before the glacis, ſuch as detached redouts or arrows, the Beſiegers will endeavour to get poſſeſſion of them, either by ſword in hand, or by ſaps; to prevent their being taken ſword in hand, they ſhould be well paliſſaded; and the ſaps may be retarded, by throwing a great number of grenades and other fire-works into them, alſo by ſudden ſallies, and retreating quickly; and there ſhould be ſome with ſap-hooks to overturn their gabions, others to burn them with ſome combuſtibles; in ſhort, all means ſhould be uſed, that art or cunning can deviſe, or time and ſituation allow.

If there are any detached redouts at a ſmall diſtance from the covert-way, the Beſiegers will carry their ſaps round them, to take them by the gorge; thoſe who defend ſuch works, after they have made all the defence poſſible, ſhould take care to retire betimes, otherwiſe they will be in danger of being taken.

The Beſieged ſhould never defend any works with too much obſtinacy, by which they may loſe many men to little purpoſe, as being always overpowered by a ſuperior number, but on the contrary, to retire in proper time; and ſo ſoon as the enemy are fallen to work at their lodgments, to ſecure their new poſſeſſion, a ſudden and brisk ſally ſhould be made, to drive them out, before they have time to ſettle, and deſtroy their works; this

ſhould

ſhould be repeated ſo oft, as it may be done without too great a loſs of men, and may ſucceed for ſome time, as being near the covert-way, where ſallies may be ſo well ſupported from thence, as to make the foremoſt of the enemy give way, and theſe never fail to throw the reſt into confuſion; in the mean time the works are poſſeſſed, which were before abandoned; but the Beſieged will at laſt be obliged to give way to ſuperiority, without any hopes of returning; then it is that the Mines, which ought to have been placed under theſe works, ſhould be ſprung; this will not fail to deſtroy their lodgments only, but themſelves likewiſe, if they are well executed. This may, perhaps, afford the Beſieged an opportunity to return once more, whilſt the enemy are yet in confuſion.

The enemy being now in poſſeſſion of theſe out-works, will endeavour to render the caponiere which leads to them, uſeleſs, either by enfilading them with their guns, or by rolling ſtuffed gabions before them; this proceeding of theirs muſt be oppoſed by the fire of the great and ſmall arms, by ſhells, ſtones and grenades, as by ſallies, likewiſe; all which may retard though not ſtop their courſe.

At the ſame time that a Place is fortified, a gallery with ſtone or brick ſhould be made all round the Counterſcarp of the covert-way; and from this others under the ſaliant angles of the glacis and the places of arms in the re-entring angles, nearly as far as the glacis reaches: this may be done at an eaſy rate. For inſtead of making counterforts, arches are made as at *Bergen-op-zoom*, which, I have ſhewn in my practical fortification, is very little more expenſive, and when the Place is attacked, branches

branches are carried to the right and left, ſo as to make mines under every part of the glacis; and as faſt as ſome are ſprung, others are made ready; this will oblige the Beſiegers to proceed with great caution, and to ſecure their works above-ground, muſt carry galleries and make alſo mines under-ground, whereby the Siege will be conſiderably retarded.

It is in this place, ſays the famous miner Mr. *de Valliere*, where the greateſt uſe of mines may be made; for if they are properly contrived, every part of the glacis may be blown up ſeveral times; and if this be well executed, will reduce the Beſiegers either to abandon the Siege, or if they are bent upon the taking the Place, make them pay dear for it.

The manner how theſe mines are to be placed and executed, will be ſhewn hereafter in the treatiſe of mines.

The enemy being come near the covert-way, will prepare to attac it, either by ſword in hand, or by means of ſaps, which they carry along the ridge of the glacis. To prevent the covert-way from being inſulted ſword in hand, a double row of paliſſades ſhould be placed in it, at four or five feet diſtance from one another, with crows-feet placed between and before them; likewiſe ſacks full of powder, to ſet on fire when the ground can no longer be maintained; coffers of three feet long and a foot ſquare are alſo placed three feet under ground, filled with ſhells and grenades.

In the mean time, frequent ſallies ſhould be made, either to deſtroy their materials, if poſſible, or to retard their Approaches. But if they continue their Approaches, they ſhould be oppoſed

with

with vigour and force; till overpowered at laſt by ſuperior numbers, it will then be proper to retire behind the traverſes at the ſaliant angles, and into the redouts of the places of arms, making, at the ſame time, a general diſcharge of great and ſmall arms, from all the works facing the attac, and ſetting fire to the ſacks of powder, coffers and ſhells, which were placed under the glacis.

If this ſhould happen to throw the enemy into much diſorder, they muſt be ſuddenly attact to drive them out; and if that fails, fired upon from all the adjacent works.

This kind of Defence having been continued till the enemy is ſo well eſtabliſhed in the covert-way, as that it is impoſſible to drive them any more out of it, the Mines are ſprung, to deſtroy their lodgments.

The Beſiegers having got poſſeſſion of the covert-way, will prepare to attac the redouts in the places of arms; but if they have revetements, and their ditch a good row of paliſſades in the middle, they will give the enemy no little trouble, and oblige them to make their Attacs by ſaps and Mines; this the Defenders will oppoſe, with all that art and ſtratagem can invent; and when they have done every thing in their power, and the enemy are near enough, and juſt ready to ſtorm, they will at laſt retire, ſetting fire to the Mines, to blow up the redouts, together with the works near them. Great care muſt be taken, in the ſpringing Mines in the covert-way, that they may not throw the Counterſcarp into the ditch, which would ſave the Beſiegers the trouble to make a deſcent.

If the ditch of the Place is wet, ſmall Floats are made, with parapets on each ſide, musket-proof,

proof, which ſerve to bring off troops from the redouts, and other out-works, when they can no longer be defended, as alſo to diſturb the enemy in the covert-way, which they approach in the dark, give a diſcharge into the works, throw grenades in them, and overturn the gabions with ſap-hooks and then retire, to let the fire of the oppoſite works act, where the workmen being uncovered will do much damage and retard the work. When the deſcent into the ditch is making, the Beſieged approach near the opening into the ditch by means of theſe Floats, ſet fire to the faſcines which are above water, and fire upon the workmen: and laſtly, to deſtroy the enemy's miners who are ſent to lodge in the breach.

It is to be expected that the enemy will give ſeveral falſe alarms, as if they intend to aſſault the covert-way, to put the Beſiegers upon ſpringing their Mines, which ſhould however not be done till it can be deferred no longer; then thoſe moſt advanced are fired firſt, ſo as not to render the others uſeleſs; and if the Beſiegers make Mines likewiſe, they ſhould be ſearched for, either to deſtroy or render them uſeleſs; but if that is not to be done without the hazard of being diſcovered or blown up, it would not be improper to make a ſhew of an intended ſally, before the Mines are ſprung, to draw great numbers of the enemy towards the place and increaſe their effect.

If the garriſon is too weak to defend the covert-way with open force, a ſmall party left in the ſaliant angles only, with orders to give a general diſcharge, and retire ſo ſoon as the enemy are advanced within ſome paces of the paliſſades, ſetting fire to the coffers, ſhells and grenades buried under the glacis; and

and ſo ſoon as the enemy are got within the covert-way, a general diſcharge is to be made with all the fire-arms from every part of the oppoſite works, and immediately after a ſtrong ſally ſhould be made from the places of arms to recover the covert-way; and this Defence is to be continued till ſuch time as the enemy have quite finiſhed their lodgments.

If the Beſiegers approach the covert-way by ſaps, and make no galleries for Mines, ſallies are ſometimes feigned, and ſometimes executed; ſhells and grenades are thrown to retard the work, and when their lodgments on the ridge of the glacis are made and filled with troops, their Batteries finiſhed and ready to fire, the Mines are ſprung, which if well executed, ought to blow up both their lodgments and Batteries at once; but if the enemy have any galleries by which they find out the mines, it will be neceſſary to deſtroy their cavaliers by guns placed in the works oppoſite to them, and by ſhells and ſtones thrown into them; and a general ſally ſhould be made on all their works at once.

The DEFENCE *of the* PASSAGE *over the Ditch before the Ravelin.*

THE Beſiegers having got poſſeſſion of the covert-way, and perfected their lodgments in it, will fall to work at their Batteries for making breach, and at the deſcent of the ditch; this work muſt be oppoſed by the fire of both great and ſmall arms; if the ditch is dry, ladders may be ſet againſt the counterſcarp, and from them grenades thrown into the enemy's lodgments, while ſome with ſap-hooks

hooks pull down the gabions, ſand-bags, *&c.* to uncover and expoſe the workmen to the fire of the Place. Mines may be uſed here with great advantage, for they furniſh various means to fatigue the Beſiegers and obſtruct their work; to make them loſe likewiſe much time and many men.

Mines ſhould particularly be placed under the Batteries of the Beſiegers to blow them up, which may be done ſeveral times, if the ground permits, to make ſeveral under one another *, by which the enemy will loſe much time in the repairing them; and if theſe mines are rightly placed, they may throw the guns every time into the ditch, by which they will be obliged to bring others from diſtant places, and will never be ſecure till they are maſters under-ground; and while they are repairing their works, the retrenchments in the ravelin and baſtions may be finiſhed, if not done before.

Whilſt the enemy are preparing their Batteries to make Breach, they work at the deſcent into the ditch at the ſame time; which, when made under-ground, the miners ſhould endeavour to oppoſe them; if the ditch is dry, ſmall detachments of five or ſix men may be placed near the counterſcarp, to watch the moment when the enemy break through

* At the Siege of *Turin,* in 1706, the Beſieged blew up the Batteries of the Beſiegers which were on the covert-way ſeveral times; by this, and other ſtratagems, they gave time to prince *Eugene* to come up, and drive the *French* out of their Lines, and relieved the Town.

Mr. *Belidor* made three Mines at *La Ferre* in 1729, one under another, under the Batteries which were to make Breach in the Lunets, at the ſide of the ravelin, with which we blew up theſe Batteries three ſeveral times, and every time tumbled the guns into the ditch.

through the revetement of the ditch to fire into the gallery, and then retire on each ſide of the opening to load their arms again; this may be repeated ſeveral times, whereby the ſappers will either be killed or obliged to retire, and which muſt retard the works; fire-balls and grenades may likewiſe be thrown into this opening with good ſucceſs on this occaſion.

Beſides what has been already mentioned concerning the Defence of a dry ditch, caponiers are made in the entrance near the ditch of the Place; that is, a kind of parapet ſloping like a glacis croſs the entrance of the ditch before the ravelin, behind which a few ſmall guns are placed, whoſe fire, together with that of the oppoſite faces of the baſtions, well directed, will increaſe the Defence conſiderably. Theſe caponiers ſerve alſo to ſally from thence, and to fall on the workmen in the ditch, which may be done with ſafety, by reaſon of the ſecure retreat.

It is certain that the paſſage over a dry ditch may be diſputed with great advantage for a conſiderable time, by frequent ſallies, Mines and fireworks; whereas that over a wet one can only be oppoſed by the fire of the great and ſmall guns, and burning the faſcines above water; or if the water has a ſtrong current, by opening ſome ſluices on a ſudden to carry away their faſcines, and whatever elſe is uſed to make the paſſage. But as in time the Beſiegers will ſurmount all thoſe obſtacles by their ſuperiority, we ſhall ſuppoſe that they have carried their works quite up to the Breach.

DEFENCE

DEFENCE *of the* RAVELIN.

WHILE the paſſage over the ditch is making, all proper precautions are taken to oppoſe the enemy, and to defend the Breach, with all the ſtratagems and force that art and opportunity can ſuggeſt; guns are placed every where from whence the Breach may be ſeen; the retrenchments in the ravelin are put in good condition; a row of paliſſades are placed along the middle of the ditch, from one end to the other, with a ſally-port at each end, as likewiſe another on the outward ſlope of the parapet, in a horizontal manner, rather inclining a little with the points downwards, ſo that the grenades thrown upon them may roll down into the ditch. The paliſſades in the ditch ſerve as a kind of barrier for the Beſieged to retire behind, when hard preſſed, and for keeping troops there in readineſs, either to ſupport thoſe who defend the Breach, or ſecure their retreat; as likewiſe to prevent the retrenchments from being inſulted ſword in hand.

If the garriſon is ſufficiently ſtrong, two or three retrenchments may be made, with parapets of about 12 or 15 feet thick, a rampart of 15, and a ditch of 18 or 20 wide; each of the ditches ſhould be paliſſaded, as well as each parapet, in the ſame manner as was directed for the firſt.

As the ravelins are very ſpacious, a Governor who has every thing neceſſary for a good Defence ſhould not content himſelf with the bare works of the Place, not with one ſingle retrenchment in each work, but on the contrary, ſhould make ſo many as the works will admit of; which being

well

well made, and ſecured from being inſulted ſword in hand, will enable a skilful commander to defend every inch of the work, with great advantage to himſelf, and to the no leſs detriment of the Beſiegers.

If the enemy intend to take the ravelin by ſtorm, and their troops are ready to mount the Breach, a great quantity of all kinds of combuſtibles are thrown at the foot of the Breach, to be fired, and conſtantly ſupplied with fewel, when they are mounting the Breach, a great number of grenades are thrown down amongſt them, ſacks filled with powder, glaſs or earthen bottles filled with powder, and burning matches twiſted round them, thundering barrels, crows-feet, harrows, and large timbers with long iron ſpikes, faſtened with chains, ſo as not to be removed by them, as alſo ſhells ſo confined as not to roll out of the Breach; a great quantity of looſe powder is alſo ſcattered about, and fire ſet to it, at the approach of the enemy; in ſhort, all kinds of fire-works and engines made for military uſe, ſhould be employed to retard the loſs of the work. This being well executed, and fire-arms, both great and ſmall, well ſerved, together with the mortars for throwing ſhells and ſtones, will do them infinite damage, and retard the taking of the work for ſeveral days; as was the caſe of the Siege of *Turin*.

When all the Defence is exhauſted, and the enemy got at laſt poſſeſſion of the Breach, the Mines are fired, to deſtroy, if poſſible, all thoſe in the Breach, and overſet all their works in it; which being perceived by the Beſieged, they muſt return immediately, and clear the Breach, ſo as to render it impracticable. Several Mines ſhould be made one under another, if the ground will allow

of

of them, that the Breach may be deſtroyed ſeveral times, which will be a means to diſhearten the enemy's troops, and put them quite out of patience.

If they ſhould attempt to get poſſeſſion of the Breach by means of ſaps, they muſt be continually haraſſed with fire-works of all kinds; grenades, even ſhells, may be rolled down amongſt them, and they ſhould frequently be ſuddenly fallen upon, by a party of ten or twelve men, to drive them out and deſtroy their works; and ſometimes ſmall Mines may be ſprung; and laſtly, when all endeavours, either by force or ſtratagem, are overcome by the obſtinacy and ſuperiority of the Beſiegers, as the laſt reſource of the Beſieged, a great Mine ſhould be ſprung, which being well executed, will deſtroy the Breach in ſuch a manner as to make it abſolutely impracticable.

But ſuppoſe, the Beſiegers are maſters of the Breach, and their lodgments in it ſo well ſecured, as to make it impoſſible for the Beſieged to drive them out, all poſſible precautions are taken to perfect the retrenchments in the ravelin; the Beſiegers will therefore be obliged to make uſe either of Mines or guns to deſtroy them. In the firſt caſe, the miners cannot approach the parapet before the paliſſades in the ditch are broke; which when attempted by the enemy, they muſt be briskly fired upon, and a great quantity of grenades thrown in amongſt them, to give them ſome trouble in doing it; however, as it muſt be done at laſt, we ſhall ſuppoſe that they have accompliſhed it one way or other; ſo ſoon as they are broke, great quantities of fire-works, with all kinds of combuſtibles, are

thrown

thrown into the ditch, by which means the Approach will be retarded ſo long as the fire laſts.

In the mean time, a gallery ſhould be made from the ditch behind the firſt retrenchment, and carried under its parapet to lie in wait there for the enemy's miners, and oppoſe their making of Mines to blow up the firſt retrenchment; and ſo many Mines ſhould be made by the Beſieged as may blow up the whole retrenchment, when it can be defended no longer.

When a Governor is reſolved on a long Defence, and is provided with every neceſſary for that purpoſe, numberleſs ſtratagems may be uſed, both above and under ground, to fatigue the enemy, and make them loſe more men and time than the conqueſt is worth.

If the Beſiegers are put to the neceſſity of making Breach in theſe retrenchments with their cannon, it will coſt them great time and trouble to make a paſſage for their guns through the ditch, and to bring them up the Breach, as alſo to erect a Battery in the ravelin, under the near fire of the retrenchment, which cannot be ſilenced till that Battery is made; and they may likewiſe be all that time oppoſed by ſudden ſallies, Mines, grenades, and all kinds of fire-works, and at laſt when their Battery is finiſhed, and ready to fire, it may be blown up.

Whatever has been ſaid in regard to the Defence of the firſt retrenchment, may be put in practice in the Defence of the ſecond and third likewiſe, if there are ſo many; ſo that a ravelin or any other conſiderable out-work, when defended as it may be, will ſtop the enemy for a conſiderable time.

This

This has feldom been the cafe, if there is one retrenchment in a work, it is generally thought fufficient; but it happens oftener that there is none at all; fo that one may fuppofe, a Defence like that we have been explaining is only chimerical, if the Sieges of *Vienna* and *Candy*, both by the *Turks*, were not inftances of the contrary, where there was hardly an inch of ground either within or without thofe two towns, as far as the extremities of the glacis, and even beyond them, but what was retrenched and countermined.

The DEFENCE *of the* PASSAGE *over the ditch before the* BASTION.

WHILST the enemy are carrying on the Attacs of the ravelin, they work at the defcent and paffage over the ditch before the baftion at the fame time; all that has been faid with regard to the defence of the ditch before the ravelin, may be executed in the defence of this: to which may be added, that the caponier leading from the middle of the curtain to the ravelin, when the ditch is dry, will be of great ufe to fire upon the enemy in their paffage, and to fally from thence, to deftroy their works: a row of paliffades may be placed in the middle of the ditch, before and parallel to the faces, and at each end of them, oppofite to the flanks, caponiers or breaft-works may be made fomething lower than thofe before, and fo as to flank the paliffades, behind which fmall guns may be placed loaded with grape-fhot to defend the paliffades.

All

All caponiers or retrenchments made in the ditch, ſhould be covered with ſtrong planks, to ſecure thoſe placed there from the grenades and other fireworks, as likewiſe ſtones thrown by the enemy placed near the oppoſite counterſcarp.

But as this covering may be ſet a fire, the planks muſt be covered with ſome earth or dung, or with freſh ox-hides.

Theſe lodgments in the ditch ſhould alſo be countermined, if it may be done, or elſe coffers and ſhells ſunk in them to blow up the enemy when the works can no longer be defended; the horſe may be uſed with ſafety in the ſallies from hence, having the reſt of the ditch to retire into.

If the ditch is wet, and has tenailles, the fire from them will very much obſtruct the paſſage; and the floats or boats mentioned before, may approach the Beſiegers works in the dark, pull down their parapets of faſcines with their long iron hooks and forks, and thereby expoſe the workmen to the fire from the baſtions, curtains and flanks; and when their works are raiſed above water, a quantity of fire-balls and other fire-works are thrown upon the faſcines to burn them.

If there are any dykes or ſluices, they may be opened ſo ſoon as the work is raiſed to any height, that the rapidity of the current ariſing from thence may overſet and carry away the works. But it muſt be obſerved, that if there is a ſluice to let in the water it ſhould never be opened till the laſt extremity; for if once the ditch is filled all the ſtratagems, which a dry ditch can afford, are over; on the contrary, when all the defence is exhauſted, and the Beſiegers are mounting the Breach, if then the water is let in, it will put them into ſuch confuſion, that

that they may be eaſily repulſed with the loſs of all thoſe that have paſſed the ditch, provided the water can be raiſed ſuddenly to a great height.

When *Landau* was beſieged by the *Imperialiſts*, Mr. *de Valliere* adviſed the Governor not to open the ſluices till the aſſault was made, and by not taking this advice, the Place was taken much ſooner than it would have been, had the ditch not been filled ſo ſoon.

Mr. *Belidor* in his *Architecture Hydrauliques*, has ſhewn, that if proper ſluices were built under the bridges, and at the ends of the ditches, where a ſufficient quantity of water can be had, how one ditch may be kept full of water whilſt all the others are dry, and how to fill or empty the ditches as occaſion requires. If this ſcheme was uſed, it would be next to impoſſible to take the Place by open force, at leaſt not without loſing more men, and be at greater expences than the Place is worth. The defence before the baſtion is certainly the moſt ſhining part of a prudent Governor; without expoſing the garriſon to be cut off, or even going out of their works, he may put all ſtratagems in practice to deſtroy the enemy's works and workmen many times; while, on the contrary, the enemy can oppoſe him but with a ſmall number, and which can only be ſupported from places at a great diſtance.

The DEFENCE *of the* BASTION.

THE Beſiegers having finiſhed the paſſage over the ditch, and the Breach ready for an aſſault, they muſt be oppoſed by the ſtouteſt men of the garriſon; ſome of them ſhould be armed with breaſt-

breaſt-plates on this occaſion, and long pikes in their hands to defend themſelves, and to repel the enemy if poſſible. It is ſuppoſed that a Governor who is reſolved to defend the Breach in the baſtion has made one or more retrenchments in it, and placed a row of paliſades in the middle of the ditches before them; and if theſe retrenchments are made with faces and flanks parallel to thoſe of the baſtion, Mr. *Vauban* would have rows of paliſades croſs the rampart, for the defenders to retire behind when drove back by the aſſailants; not only ſmall light guns ſhould be placed in theſe retrenchments but likewiſe mortars to fling ſhells and ſtones, by which, when the Beſiegers are lodged in the Breach, they may be ſo haraſſed as to be unable to keep that poſt. Mines ſhould likewiſe be made under theſe works, ſo that when the Breach is loſt, the Beſiegers cannot come underground to blow them up, and to oblige them, if poſſible, to make a battery in the Breach to batter theſe retrenchments, which will coſt great trouble, and during that time other ſtratagems may be uſed to retard the work. The ſeveral ſtratagems mentioned in the defence of the Breach of the ravelin are to be practiced here; and when the defenders are obliged to give way to the ſuperior force of their enemy, they ſhould open to the right and left, and retire along the parapets, to leave room for the guns placed in the retrenchment to give a general diſcharge upon the enemy, which ſhould be done with grape-ſhot; if this is well executed in proper time, it will ſurprize, if not drive the enemy back; which being perceived, a vigorous and ſudden onſet ſhould be made to puſh

them

them down the Breach, if poſſible, and then to repair it.

Many other ſtratagems may be uſed in the defence of the Breach, beſides thoſe mentioned; as for example, water-engines may be placed in the retrenchments with leathern pipes, which would make a kind of artificial rain, and being directed towards the Breach ſpoil the enemy's fire-arms, ſo as to make them uſeleſs for the preſent; engines may be contrived to fling heavy ſtones, or large pieces of timber amongſt them, or to blow duſt or ſand in their faces; in ſhort, a man of invention may find a thouſand oddities to ſtop an enemy upon ſuch occaſions.

When *Alexander* the great beſieged *Tyre*, the Beſieged filled armours with ſand, which they heated, and threw them among the Aſſailants, whilſt they were endeavouring to climb the wall; the ſand, thus heated, ſtuck ſo to their cloaths that it was impoſſible to get it off, and burnt them to the very bones.

Hannibal, the *Carthaginian*, in a ſea engagement, threw pots filled with ſnakes amongſt the enemy; the pots being broke the ſnakes run about all over the ſhips, which terrify'd them to ſuch a degree that he preſently conquered them.

Whoever reads of the ancients manner of attacking and defending Places, will find that they were much ſuperior to the moderns in their contrivances; and ſince the invention of gun-powder all thoſe engines which were of uſe before have been entirely neglected; but I am perſuaded that many of them would be as uſeful now as ever, eſpecially in the defence of Places.

When

When all ſtratagems are exhauſted, and the enemy at laſt ſucceed, the Beſieged retire and ſet fire to their Mines, as in the defence of the Ravelin, which being well executed, will deſtroy the Breach ſo effectually, as to make it impoſſible for the Beſiegers to lodge in it, as happened at *Maeſtrich* when it was beſieged by King *William*, then Prince of *Orange*, and defended by *Calvo* the *French* Governor; the Beſiegers having lodged in the Breach at the ſaliant angle of the baſtion were blown up by a Mine, and the Breach deſtroyed, ſo as to render it impracticable to repair. They made another Breach near the angle of the ſhoulder, and when they had lodged in it, were blown up as before; they then made one near the other angle of the ſhoulder, where they were ſerved in the ſame manner; by this means the flower of the Army were deſtroy'd, and Prince *Conde* coming up with a ſmall Army to relieve the Town, the Siege was raiſed.

But ſuppoſing the Mines ſhould not produce the entire effect that may be expected, they will at leaſt diſorder the Beſiegers ſo much, that if they are charged with vigour and ſuddenly, they may be drove out and the Breach repaired.

If the enemy attempt to lodge in the Breach by means of ſaps, they muſt be charged often with ſmall parties of ten or twelve men only; ſhells, ſtones, and grenades thrown amongſt them, as in the Defence of the ravelin, and when it can no longer be defended, the Mines are ſprung. If the ditch is dry, the enemy may be diverted from their deſign, by Mines carried on underneath the ravelin from the gallery, which ſerves as a communication betwixt the ravelin and the curtain; ſo ſoon as the enemy prepare to aſſault the baſtion, fire is ſet to theſe

theſe Mines to blow up their lodgments in the ravelin, and which being immediately taken poſſeſſion of by the Beſieged, they may fire from thence upon the Beſiegers in the Breach of the baſtion; this will oblige them to retake thoſe lodgments, and in the mean time the Breach in the baſtion may be repaired.

If at the ſame time another Mine is carried under the Breach of the ravelin, and ſo ſucceſsfully ſprung, as that the enemy's works there are deſtroyed, and themſelves driven out; they cannot proceed in their attac, till they have taken the ravelin, which cannot be done without great trouble and loſs of time, beſides diſmaying their troops, ſeeing all their labour loſt, and dangers returned: ſuch a maſter-ſtroke may not improbably occaſion the raiſing the Siege, or reduce the enemy ſo low, as to make it eaſy for the Beſieged to relieve the Place.

While the enemy are buſy in retaking the ravelin, the Governor ſhould lower the flanks which were ruined, and make new embraſures; they ſhould not be quite opened till the guns are ready to fire; and if the Governor adds another flank within the baſtion or Battery higher than the other, he will be in a condition to diſmount the enemy's guns oppoſite to theſe flanks, having more to oppoſe them with than they can bring againſt him.

The enemy having retaken the ravelin, and repaired the diſordered paſſage over the great ditch, as likewiſe the Breach, will attac the baſtion again, which is therefore to be defended with the ſame force and ſtratagems as before; and when the Beſieged can maintain their ground no longer, retire behind the paliſſades in the ditch of the retrenchment,

ment, ſetting fire to their Mines, if there are any left, whilſt thoſe within the retrenchment fire with all their might into the Breach with guns and mortars, ſtones, grenades, and ſmall arms.

The enemy having made themſelves maſters of the Breach, and finiſhed their lodgments in it, will be obliged to take the paliſſades and retrenchments with the ſame trouble and precaution as thoſe in the ravelin; they muſt therefore be defended in the ſame manner as thoſe, and this muſt be continued and repeated till the Beſieged are reduced to the laſt retrenchment, when, the Governor having no more ground left to defend, may agree to a capitulation, which, if it is not very advantagious, cannot but be very glorious, both for him and all thoſe who fought under him.

Tho' the uſual method is to attac the baſtions, yet it has happened that the enemy have attact the curtain, being aware that the baſtions were well retrenched, as Prince *Eugene* did at *Liſle*, and therefore the Beſieged ſhould prepare againſt ſuch an attac, by making retrenchments at each end of the curtain, croſs the rampart, to flank the Breach, and placing guns in them loaded with grape-ſhot; they ſhould likewiſe pierce the walls of the buildings oppoſite and near the Breach, for loop-holes to fire through, and place troops in all the adjacent places, wherever they can have a view of the enemy in the Breach; as the Inhabitants of *Barcelona* did, where the *French* and *Spaniards* together were driven twice from their lodgments on the rampart, and at laſt allowed terms of capitulation.

Of

Of RETRENCHMENTS, *and their Figure.*

RETrenchments ſeem to have been much more in uſe formerly than at preſent, though it is impoſſible to tell why they are neglected; they are now uſed but ſeldom in any works; the Governors content themſelves with defending the works of the Fortification only, without taking any farther trouble to make new ones, whereby the Siege may be lengthened; and it has generally happened of late, that ſo ſoon as the Beſiegers get maſters of the covert-way, the garriſons have capitulated. It is moſt certain, that for ſuch defences it were ſuperfluous to make any additional works, ſince the chief of thoſe already made ſeem of no farther uſe than to make (as term'd) an honourable capitulation; but if the body of the Place, and the large ditch round it, are to be of no more uſe than they make of them at preſent, it would be as well to make only a ſingle rampart, without either baſtions or curtain, and a ſmall ditch; this would be ſaving of expences, and yet produce the ſame effect.

The intention of this work is to point out ſuch means as may conduce to a long and vigorous defence, it will therefore not be improper to ſhew the advantages of retrenchments, when they are well conſtructed and defended by a skillful Governor.

As their parapets are to be ſomewhat lower than thoſe of the works, in which they are made, they can be ſeen from no other place than from within the works; ſo that if the enemy intend to batter them, they muſt make a paſſage for their cannon over the ditch and through the Breach, which

cannot

cannot be done without trouble and loſs of time. Their uſefulneſs conſiſts in this, that they afford a ſafe retreat to thoſe who defend the Breach, enabling them to reſiſt to the laſt extremity, without any danger of being made priſoners; whereas, on the contrary, when there are no retrenchments, the troops having nothing to depend on but the mercy of the enemy, grow diſheartened, and oft retire ſo ſoon as they are vigorouſly attacked.

When the Beſiegers are in poſſeſſion of the Breach, the guns of the retrenchments loaded with grape-ſhot, together with their ſmall arms, do very great execution amongſt them, and oft drives them out. Nor can they be ſafe in their lodgments, ſo long as they are not maſters of the retrenchment; for the Beſieged may ſally from thence whenever they pleaſe, fall upon them, and deſtroy their works.

They likewiſe obſtruct the progreſs of the enemy's miners, who cannot get at them till the paliſſades in the ditch are broke, whereas thoſe of the Beſieged begin their galleries at the ſame time as the Attac, and have no more to do but to wait for their opponents, watching which way they come, and, when they are near enough, to deſtroy them.

This, with what has been already ſaid with regard to their Defence in the ravelin, will ſufficiently convince the reader of their uſefulneſs; it remains now to ſay ſomething of their conſtruction and ſituation. They may be placed in all kind of works, where there is a ſufficient ſpace for them, as in baſtions, ravelins, horn and crown-works. Their figures are various, ſome in the ſame form of the work itſelf, others in that of the front of a polygon, that is, with two half baſtions and a curtain; and ſometimes with one or more rentring angles

angles or bendings. *Plate* XXI. repreſents ſeveral kinds.

The DEFENCE *of a* HORN *or* CROWN-WORK.

THESE kind of works are or ought to be made only for incloſing a ſuburb, or ſome ſpot of ground, which may be advantageous to an enemy, and which falls without the compaſs of the body of the Place. When they ſerve to incloſe any building, their Defence is exactly the ſame with that of the body of the Place, with this difference only, that they may be defended to the laſt extremity without endangering the Place. When the inſide is not taken up with any principal buildings, having only magazines, ſtore-houſes, or hoſpitals there, which is generally the caſe in new-built Places, thoſe buildings muſt be demoliſhed; and as their inſides are very ſpacious, they may admit of ſeveral retrenchments, ſome of which at leaſt ſhould be made in time of Peace, with revetements, and wet ditches before them; which muſt be enfiladed by the ravelin, and when the Place is inveſted, and the Attac made on that ſide, more may be made. It is certain, that an experienced Governor, who has every thing in plenty for his Defence, may defend a work of this nature a long time, and diſpute the inſide inch by inch; and when the enemy have got poſſeſſion of one retrenchment, may drive them out from another, either ſallying or blowing them up with Mines.

Theſe kind of works have not always made that Defence, which might have been expected from works ſo conſiderable, and ſo expenſive in their conſtruction; but, in my opinion, it has rather been owing

owing to the inabilities of the Governors, or the want of neceſſaries to defend them, than to any defect of the works.

There is nothing more particular in the Defence of theſe retrenchments, than thoſe of the ravelin, which I have explained; it is only to be obſerved, that M. *Vauban* allows no more troops for a horn or crown-work than for a baſtion; which is not ſufficient for a good Defence; there ought to be at leaſt twice that number, that is, about 2000 men.

As horn and crown works preſent but a ſmall front to the Beſiegers, and which cannot be ſeen or defended by any other work, their reſiſtance cannot be very great; but as they are very ſpacious within, retrenchments upon retrenchments may be made; and if their ditches have one or two rows of paliſades, and theſe works countermined, there is no doubt that the inſide of theſe horn or crown-works may defended inch by inch to the very laſt, with all poſſible advantage. For the Beſiegers muſt make a Breach into every one of the retrenchments, either with cannon or by mines: if they make uſe of the former, it will coſt them much trouble and time to erect the batteries, and bring the cannon over the ditches, where the Beſieged are ſo near, and can ſally upon them, nail up their cannon almoſt when they pleaſe, and at laſt blow the batteries up by mines. If they make the Breaches by mines, they muſt carry their galleries under the ditches and paliſades; and as the Beſieged's miners lay in wait, may deſtroy them whenever they come near; all that the Beſiegers can do, is to make Breaches in the branches as well as in the front, and endeavour to deſtroy the retrenchments with ſhells, and annoy the

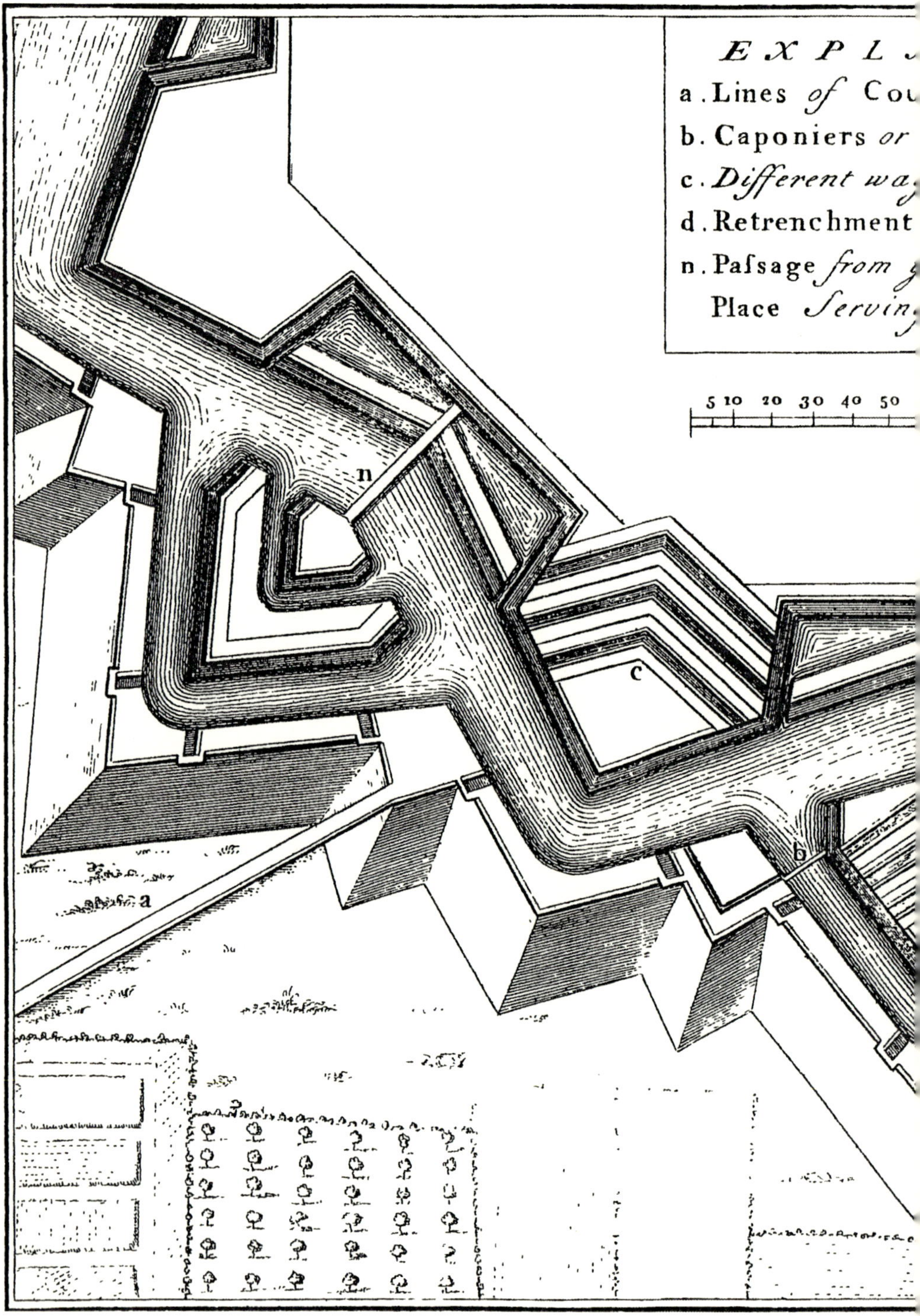
EXPL
a. Lines of Cou
b. Caponiers or
c. Different wa
d. Retrenchment
n. Passage from
Place Servin
5 10 20 30 40 50
n
c
a
b

XXI.

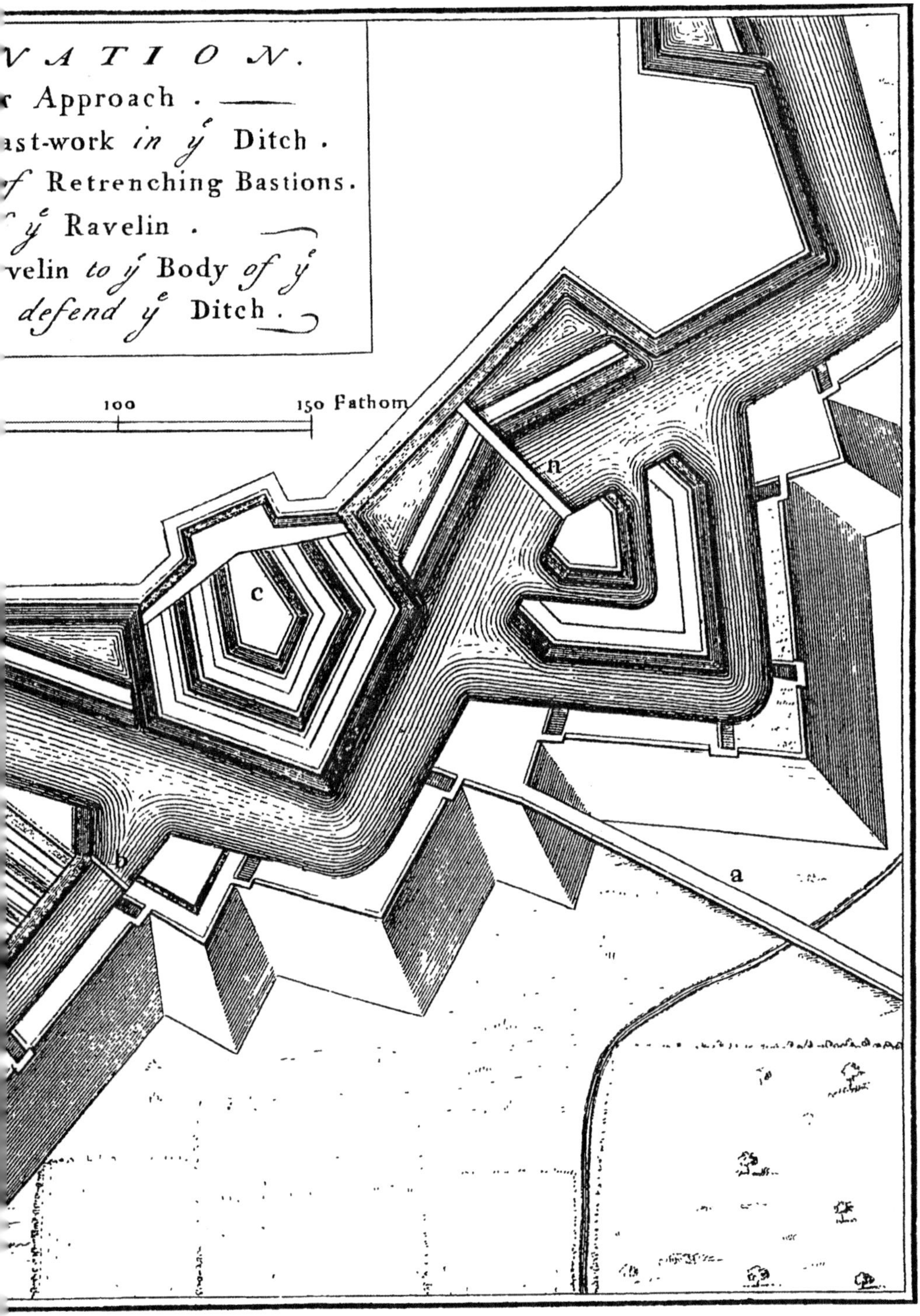

VATION.
r Approach.
ast-work in ye Ditch.
f Retrenching Bastions.
ye Ravelin.
velin to ye Body of ye
defend ye Ditch.
100
150 Fathom
n
c
b
a

the defenders with ſtones; this will require much time and loſs of a great number of men. I am perſuaded that if theſe works were defended in the manner they might, it would require at leaſt a month to take any one of them.

The DEFENCE *of a* SECOND DITCH *and a* SECOND COVERT-WAY.

SUPPOSE that the ſecond ditch is full of water, ſo high as to be even with the foot of the firſt glacis; they ſhould at leaſt be conſtructed in this manner, when the nature of the ground will admit of it. The Defence of the ſecond covert-way, and the lunets, are exactly the ſame as that of the firſt, when there is but one, and the ravelins. As the fire from the covert-way is grazing, and may be accompanied with ſtones, ſhells, and grenades, the paſſage over the ſecond ditch, as alſo the making the lodgments on the firſt glacis, will meet with great difficulties; and whenever the enemy attempt it, which cannot be done till about the middle of the glacis, for want of ſufficient depth of earth above the level of the water, ſallies are to be made to drive them back; which may be done with ſucceſs, the firſt covert-way being ſo near, and the enemy not having it in their power to ſupport their workmen, but by a few men only.

The enemy having made their lodgments on the firſt glacis, near the firſt covert-way, they muſt be diſturbed ſo much as poſſible above ground; and when that can no longer be done, without viſible danger to the Beſieged, they ſpring their Mines or coffers, if they have any. As to the reſt of the Defence,

Defence, it is the ſame as if there were no ſecond covert-way or ſecond ditch.

If the ſecond ditch has ſluices ſo as to keep it dry, or let the water in upon occaſion, if ſhould be kept dry till the Beſiegers have taken the ſecond covert-way, and have made their paſſages over the ditch; becauſe ſallies may be made with more eaſe and ſecurity: but ſo ſoon as the Beſiegers begin their lodgments upon the firſt glacis, then the water is let in, which will probably carry away the works made thro' the ditch, if there is a ſufficient current; or at leaſt they will be obliged to raiſe them; and if the water can be raiſed ſo high as to overflow the greateſt part of the firſt glacis, the Beſiegers have no other means left, than either to deſtroy the ſluices, or to cut a drain croſs the ſecond covert-way and glacis; but whilſt either is doing, the Beſieged are not to be idle ſpectators, but on the contrary, uſe all the ſtratagems that art and the ſituation can afford to retard it: if there are good ſtone redouts in the places of arms, and coffers full of powder lodged in the glacis, as likewiſe mines as low as can be, the defence may be carried on a great while, and nothing but ſtones and ſhells can hurt them; theſe and many other ſtratagems, ſhould be uſed to retard the taking of any of the works; for if every part of a Place is defended in the beſt manner, that of the whole muſt be honourable to the Governor and garriſon.

The DEFENCE *of* LUNETS, TENAILLONS, *and* COUNTER-GUARDS.

LUNETS and Tenaillons are capable of great Defence, on account of their being ſo near the ravelin

ravelin by which they are defended, and from whence the Befieged may fally feveral times to retake thefe works when loft; the prodigious defence which the Tenaillons made at *Lifle* fufficiently proves their utility, the Befiegers having been driven out three feveral times before they could entirely take them.

The Befiegers are never fecure in thefe works till they are mafters of the ravelin; and if the Tenaillons have retrenchments in them, as cuftomary, fo well as the ravelin, the enemy can never get poffeffion of them but with the utmoft difficulty, and the lofs of many men. A brave Governor may make the Befiegers pay dear for taking thefe works, and prolong the Siege confiderably.

It may be obferved, that the enemy muft bring guns into the Tenaillons to batter the ravelin in Breach, and if the Befieged have any Mines placed fo low as water-mark to blow up thefe Batteries, it will be difficult for the Befiegers to make new ones, efpecially if the Mines are conftructed, fo as to throw the guns into the ditch of the ravelin, and when there are guns in the retrenchments of the Tenaillons.

Counterguards before baftions are works of an excellent conftruction, and may be defended long, becaufe the baftions being fo near, they may from thence pour grape-fhot, ftones, and fhells into the Breach, fo foon as the enemy appear, and may prevent their making a lodgment there; the guns from the flanks likewife may fcower the branches of the Counterguards from one end to the other; fo, if there are retrenchments in them, as cuftomary, the Befiegers will find it difficult to take them; and

and as the flanks cannot be battered but from the Counterguards only, the bringing guns over the ditch through the Breach, will be attended with much trouble and difficulty, and may be oppofed by many ftratagems; and when the Befiegers have at length made their Batteries, they will hardly have room to place fo many guns there as the Befieged have in their flanks, efpecially if they are double: and this is what is aimed at in the conftruction of Counterguards, fince fo long as the flanks are not deftroyed, the Befiegers dare not venture on the paffage of the ditch.

Counterguards placed before the ravelin are defended in the fame manner as a ravelin with a redout within it.

REMARKS.

IF there are Counterguards before the baftions and Tenaillons joined to the ravelin, and both have retrenchments with revetements, as at *Lifle*, in the front attacked in 1705, the Befiegers will find it hard to get poffeffion of the Place, if it is well provided; fince it will be impoffible to keep poffeffion of the Breach till the retrenchments are deftroyed, which can be done no otherwife than with fhells, and during the affault to fling fuch fhowers of ftones that the Defenders cannot ftand them, and as this is the only means by which the Befiegers can expect to fucceed, the Governor fhould provide againft it before hand; the manner of doing this, before the Place is Befieged he fhould confider where the Befiegers muft place their Batteries, and place mortars on the ramparts of the Place, and in the ravelin; to know what quantity of powder

der is required to throw the ſhells into thoſe Places; which being known, when the Place is attacked, he may deſtroy the Beſiegers batteries whenever they are erected; and as this can be done behind any parapet, and there being no embraſures to diſcover their ſituation, they may do their buſineſs before they can be fired at, and then they may be removed to ſome other place, by which it will be a hard task to deſtroy them.

Tho' I have ſaid, that there ſhould be but 12 mortars in a place of ſix baſtions according to M. *Vauban*'s eſtimate, yet I image there ſhould be many more, eſpecially when a place is fortified in the manner here ſuppoſed; in my opinion they are full as uſeful, if not more than guns: ſince if they are well directed, the Beſiegers will hardly be able to keep in their works which are upon the glacis and covert-way, and their ſituation never known before their effect is felt: whereas on the contrary, guns requires plat-forms and embraſures, and cannot be ſo eaſily removed, are generally diſmounted ſo ſoon as they begin to fire, by the ſuperiority of the Beſiegers. It is ſhameful to ſee a Place, that coſt immenſe ſums to fortify, hold out no longer than they commonly do: if a Prince or a State is at ſuch vaſt expence to fortify a Place, he ought to know at the ſame time, that if it is not ſufficiently provided with every thing accordingly, he only exhauſts his treaſures, without an adequate benefit. No man ſhould be Governor of ſuch a Place, but who knows how to defend it to the laſt extremity, with all the advantages its works will admit or are capable of: but it generally happens that an old officer, who has ſerved a great while, is rewarded with ſuch a poſt, either becauſe he always behaved

well,

well, or knew how to insinuate himself into his master's favour, by a submissive attendance, without considering whether his age renders him incapable of so laborious a task, or whether he has sufficient talents to make a proper defence.

Tho' an officer may have distinguished himself in many actions; yet, if he does not understand Fortification well, and has not a perfect knowledge of the practical part of Artillery, he can never be a good Governor. The great misfortune is, that brave men depend too much upon their courage, and neglect the most essential part of the art of war; or is there any care taken to make them study it, when young; it is thought sufficient, when a gentleman is made an officer, to have courage, and to behave in a genteel manner; how far this custom prevails at present is too well known.

The DEFENCE *of* FAUSS-BRAYS.

THERE are but few Places fortify'd with these kind of works at present, besides the Citadel of *Tournay*; since the invention of the ricochet firing, the Defence of them is dangerous, and of no duration; they ought to have many traverses made in them of about 10 or 12 feet thick, and as high as the parapet, to prevent the effect of these ricochets; for which purpose it would likewise be proper to have behind each traverse a kind of shed, covered with strong planks of three or four inches thick, and roofed like a pent-house, that those who defend them may be secure against the ricochets and grenades; but as the enemy will not fail of throwing many shells into the Fauss-bray, it will be be troublesome to screen the defenders against their effects, except there is a ditch of 20 or 24 feet wide, between them and the higher rampart.

All the Defence that such works are capable of, consists in sallying from behind their traverses, to destroy the lodgments which the Besiegers are obliged to make from the lower parapet to the upper one; however, this may be done for some time, if the place does not become too dangerous from the ricochets, shells and stones, in which case it ought not to be defended with obstinacy; the troops may be employed to more advantage in the Defence of the body of the Place.

The DEFENCE *of a Place fortified with* TOWER-BASTIONS.

THE Defence of such Places differs but little from one fortify'd with bastions, only as the counterguards are detached from the body of the Place, they may be defended to the last extremity; and as they are very spacious, admit of several retrenchments, made either in the same form of the counterguards, or in that of the front of a polygon, which being well palissaded, may retard the Siege considerably; these retrenchments should likewise be undermined, that when they can no longer be defended, they may be blown up, to prevent their serving the Besiegers as epaulements against the fire of the Place.

As the bridges of communication made on each Side of the Towers are entirely covered against the Besiegers fire, and the ditch before the towers but narrow, the communication from the Place to the counterguards cannot be interrupted; so that after

after all the retrenchments, which theſe works may admit, are defended and retaken to the very laſt; the guns placed on the faces adjacent to the towers; which cannot be diſmounted by ricochet firing, will do great execution, and may deſtroy the Beſiegers lodgments in ſuch a manner as to be drove out with eaſe; this may be done, ſo long as theſe guns remain unhurt, and which can be effected no other ways than by ſhells; even then the Governor has it in his power to replace thoſe that are diſmounted. The true Defence of ſuch Places, conſiſts in that of the counterguards, and not in the Towers, as authors imagine: ſince the inner parapet in intirely covered by that of the counterguard, and cannot be enfiladed on account of the towers, as I have obſerved; and the bridges of communication quite ſecure; the inſide of the counterguards may be defended inch by inch, both above and under ground, with more advantage than any other work whatſoever: and after all this Defence, the place is not open, and the gariſon may defend the paſſage over the ditch, and the Breaches made either in the Towers, or adjacent ramparts, a conſiderable time before they are reduced to capitulate.

To prevent the caſemates from being choaked up by the rubbiſh, a party of men ſhould be ſent into the ditch upon floats, if it is wet, ſo ſoon as it is dark, to remove it; and to prevent thoſe within the caſemates from being ſtifled with the ſmoke, ſome air-machines ſhould be placed in them to drive out the ſmoak and draw in freſh air; tho' there are ſeveral ſmoak-holes to let the air out, yet they are not ſufficient to clear them, unleſs other means are uſed for that purpoſe.

When

When the Beſiegers have made a breach in the inner rampart or in the Towers, and finiſhed the paſſages over the ditch, the Governor may then conſider if he is able to ſuſtain an aſſault; if he is, he ſhould do it under pretence of making an honourable capitulation, Places have oft ſurrendered before the Breach in the body of the Place has been practicable; which ſhould be ſeverely puniſhed, to prevent ſuch treacherous cowardice for the future.

Tho' the *French* Governors are ordered by their inſtructions, to ſuſtain one aſſault at leaſt in the body of the Place, yet few have comply'd without being puniſhed: for when *Landau* was beſieged by the *Imperialiſts,* the Governor capitualted ſo ſoon, as the Beſiegers got poſſeſſion of the counterguards, tho' the Towers and inner ramparts were yet good and very little damaged. Nothing but the want of every neceſſary could excuſe this Governor, that he was not puniſhed is well known, but what reaſons he alledged for his timidity is uncertain.

The DEFENCE *of a Place acceſſible on one ſide only.*

THE firſt thing a ſkilful Governor ſhould do, is to examine carefully if thoſe ſides which are eſteemed inacceſſible are really ſo; whether by ſome means the enemy may not be able to approach that way; or if they are not inacceſſible, it may be poſſible to make them ſo; if he finds that it may, he ſhould immediately ſet the garriſon to work before the enemy is near to obſtruct them; he ſhould ſtrictly obſerve whether the enemy can place Batteries on a riſing ground ſo as to annoy ſome of the works; in which caſe, if the ground cannot be levelled, he ſhould

ſhould raiſe a redout there, to prevent an approach that way. But if theſe redouts cannot be ſupported and defended by the works of the Place, cavaliers may be made in the neareſt baſtions to command the ground. If any part of the Town is ſurrounded by marſhes, rivers, or the Sea, he ſhould conſider, if in a dry ſeaſon they may not be paſſed over, and take all proper precautions. There are many obſervations to be made on uneven ſituations, which cannot be foreſeen but by thoſe who live upon the ſpot, and it is the Governors duty frequently to examine, and conſider what may be done to render the Approaches as dangerous and difficult as poſſible.

A great advantage ariſing from ſuch a ſituation is, that the works on that ſide which is acceſſible are not to be enfiladed by the ricochet Batteries; and to be attacked in the front only; and as the inacceſſible ſides need only be guarded by a few men, the beſt part of the garriſon may be employed in the Defence of the works attact; as the troops are in no danger of being cut off from the Place, they may ſally out when they pleaſe, and oblige the enemy to begin their Approaches at a great diſtance: the Governor may alſo conſtruct ſome redouts or retrenchments beyond the glacis to keep off the enemy, taking care at the ſame time, that theſe outworks, when loſt, become not a ſhelter to the enemy, for which purpoſe they muſt either be mined, or made ſo as to be defended and enfiladed by the covert-way. As the Governor is acquainted before-hand which ſide is to be attacked, he has leiſure to prepare every thing for the enemy's reception; that whenever they appear, he may be able to put his deſigns in execution.

The

The DEFENCE *of a Place ſituated on a high Hill.*

THE Governor ſhould carefully examine the ground about the Place, and if any part of the hill is flat, and not taken in by the works, he ſhould make out-works there quite cloſe to the brow of it; if the avenues leading to the Place are hollow, and there are any places which command them, Batteries ſhould be raiſed there; and if there are other hills which command any part of the works, a redout, or ſome other work ſhould be made there, and a detachment put in it. In ſhort, whatever may favour the Approach of the enemy, ſhould either be eraſed or fortify'd, and this ſhould be done before the enemy appears; as likewiſe a ſufficient quantity of proviſion for a twelvemonth, or longer, ſhould be procured from the neighbouring country, and kept in magazines againſt the time of a Siege: but, eſpecially water ſhould be taken care of, as generally wanted in ſuch Places; for which reaſon the Governor ſhould keep the ciſterns for receiving the rain-water always in good order, uſing it with great œconomy, at leaſt when there is any danger of an Attac; and if there are any wells or pumps he ſhould not ſuffer the inhabitants to ſpoil them, or to uſe more water than is abſolutely neceſſary: whenever an enemy finds it too difficult to attac the Place in form, he generally turns the Siege into a Blockade, thinking it eaſier to ſtarve the garriſon than to reduce them by force, which however will not ſucceed, if the Governor has taken the neceſſary precautions before-hand: add to this, that in winter, or in bad weather, when the enemy cannot guard

guard all the avenues ſo ſtrictly as at other times, care ſhould be taken to make uſe of that opportunity to bring in corn, cattle, *&c.* that may be wanted. There are many circumſtances ariſing either from the nature of the ground about the Town, or from the time, ſeaſon, or opportunity, in which a skilful Governor may provide himſelf with theſe neceſſaries, in caſe the enemy, after finding all their endeavours to attac them fruitleſs, ſhould reſolve to turn the Siege into a Blockade.

The DEFENCE *of a Place ſituated near a river.*

SUCH Places being generally of great importance, on account of their trade, a Governor ſhould take more than ordinary care to preſerve it before an enemy appears; ſhould examine all the parts of the works, and put them in good order of Defence if they are not ſo: if the river runs on one ſide, and there is only a fort on the other to cover the bridge, as is uſually the caſe, he ſhould conſider whether when the water is low in a dry ſeaſon, the enemy may not paſs along the ſhore, and inſult the Place on that ſide, and take proper care to prevent them; and if a Battery can be erected on the other ſide to enfilade any part of the works, or make Breach, which often happens, in that caſe, ſome outworks ſhould be conſtructed to make the Place as ſtrong on that ſide as on any other. It is to be obſerved, that the enemy make their Approaches moſt commonly by the river-ſide, becauſe one of the flanks of their Attacs is ſecured thereby; and the artillery, ammunition, and ſtores, may be brought by water near the place where they are wanted;

wanted; and conſequently the Place ſhould be made rather ſtronger there than any where elſe; particular care muſt likewiſe be taken, that the enemy do not come down the river in boats by night, and ſurprize it that way.

If there are any iſlands near the Place, from whence a part of the works may be enfiladed or battered in Breach, as is the caſe at *Old Briſach* and *Fort Kell,* ſome works ſhould be made in thoſe iſlands, to keep the enemy out of them, which is of ſo much the more importance, as theſe works ſeen from thence cannot be enfiladed nor battered from any where elſe.

If the enemy are careleſs of their Trenches on the ſide next the river, the Governor may order ſome men over in boats in the night, if he has any, to attac and enter the Trenches on that ſide, and at the ſame time make a ſally on ſome other part of them. In ſhort, a ſkilful Governor will know how to make the beſt Defence, whatever the ſituation may be, whether it be water, marſhes, or dry land.

The DEFENCE *of ſmall Places.*

IT often happens in the courſe of a War, that parties are placed in poſts having little or no Fortifications, which ſerve either to ſecure the convoys of Ammunition and proviſion for the Army, or to guard ſome paſſes by which an enemy may enter the country, either to raiſe contribution or plunder it; to hinder parties from approaching either the general camp, or any troops from thence, which are beſieging a Place; or laſtly, to ſecure the troops in their winter quarters, for more conve-

niently

niently drawing them together on any fudden occafion.

An officer, who commands where he may be infulted at any time, fhould take care to fecure the entrance or gate, by a fmall ravelin made of earth, and turfed and well palifaded, as likewife the gate where he may fally out; and if there are any more gates, they fhould be walled up.

If the Place has either no ditch, or a bad one, the old fhould be repaired, or a new one made; and if there are a fufficient number of troops, a covert-way may be added, well palifaded; but if the Place is not worth the trouble, or there are not troops fufficient for that purpofe, care at leaft fhould be taken to prevent furprizes; the wall fhould be repaired where it is wanted, and put in order for Defence; all the principal avenues well guarded; fmall parties fhould be fent out in the day, to prevent the enemy from reconnoitering the Place; in the night, the patrole fhould continually be in motion, and no body fuffered to approach the gates upon any pretence whatever, to prevent the fixing of a petard.

If there are any round towers near the gate, as is cuftomary, loop-holes fhould be made flanking the gate, to fire upon any body who fhould attempt to approach.

When there is any apprehenfion of the gate's being broke open, either by a petard or other means, and there is no time, or a fufficient number of troops to fecure the outfide by palifades or outworks, a large pile of dung and earth mixed together fhould be thrown up againft the gate, on the infide, which will diminifh the force of the petard;

petard; and every means ſhould be uſed that can be imagined, to prevent the entrance of the enemy.

The centry placed over the gate ſhould keep a careful watch in the night, that no body approaches unperceived; and if he hears the leaſt noiſe, muſt immediately give notice to the guard, that they may be ready to oppoſe the enemy, in caſe of any attempt.

The troops ſhould continually watch in the night all round the rampart, or wall, to prevent an eſcalade, provided with long poles and forks to overſet the ladders, alſo with great ſtones and timbers to roll down upon the enemy; if they have any fireworks of pitch or tar, they ſhould be thrown into the ditch amongſt them, to burn or diſable them. If theſe and all other proper precautions are taken, the enemy will not have it in their power to ſurprize the Place, unleſs it be done by a great part of the army, and provided with cannon; then it muſt be left to the diſcretion of the Commander, whether to wait for the enemy, or retire betimes.

It happens ſometimes that a party is ſurprized in a houſe, church, or orchard, without having any time to fortify themſelves. It is impoſſible to lay down any certain rules for defending ſuch Places, depending on ſo many circumſtances; it muſt therefore be left to the capacity of the Commander. All that can be ſaid, at preſent, on this ſubject is, that if the houſe in which he is ſurprized is covered with ſtraw, or any thing elſe eaſily ſet on fire, he ſhould uncover it immediately, barricade the doors, make holes in the wall, in ſeveral places, high and low, to fire through, put ſome of his people in the garrets, to prevent them entering that way, others in all the rooms of the houſe; boards of the floors

above

above ſhould be taken up near the doors, to leave a ſpace of three or four feet wide, to ſerve as a kind of a ditch to this Fortification; and if there is not a ſufficient number to defend all the parts of the houſe, the lower apartments muſt be abandoned, and ſome placed over the doors, to fire down on thoſe who enter. If ammunition is ſcarce, they ſhould repel the enemy with their bayonets fixed; and if they find it impoſſible to maintain their poſt, they muſt retire in the night as privately as poſſible, and all in a body, taking care if they meet with any enemy's by the way, not to fire upon them, but to run them through with their bayonets, for fear of alarming the reſt of the body. In this manner the late Count *Saxe*, with a Party of 18 men only, reſiſted 800, and he and his party made a ſafe retreat.

Precautions which a Governor ſhould uſe in Time of War.

A Governor ſhould before any Enemy appears, examine the works of a Place, repair thoſe that want it, paliſſade the covert-way, if not done before, as likewiſe lay ſome horizontally on the middle of the Parapets, which have no Revetement; clear the ditches from the mud, ſee that the gates or entrances are ſecure and well defended from being broke up; keep ſtrict Diſcipline and good order in the Town, prevent the garriſon from moleſting and abuſing the Inhabitants; watch narrowly that no Traitors correſpond with the enemy to betray the Place; for which purpoſe, the Governor ſhould have ſome to whom he may truſt, who are to get into companies unſuſpected, to hear what paſſes, and give him notice. If there are any old aqueducts,

aqueducts, or under-ground paſſages, they ſhould be ſtopped, and centries placed at their entrances; if there is any river paſſing through or near the Town, parties muſt be put into boats in the night, both above and below the Place, to watch that the Enemy do not come that way. In froſty weather the Ice in the ditches ſhould be broke every day, and the ſhoals laid a-top of one another towards the Place, which make a ſlippery unpaſſable wall.

He ſhould ſend every day, parties of horſe and foot, to range in all the principal avenues, two or three miles diſtance, to prevent any ambuſh or approach of the enemy, in the night he ſhould take care that the ſeveral guards keep ſtrict duty, and watch carefully at their ſeveral poſts, not letting any one approach the walls, not even the centries, without the forms uſual in ſuch caſes. The patrole ſhould walk all night about the ſeveral poſts, to ſee that the centries do not ſleep, and that they liſten to hear the approach of the enemy, and on the leaſt noiſe of ſuſpicion, give notice to the guards, and they to the Governor.

On Fair or Market-days the Gates ſhould be ſtrictly guarded, the horſe and foot ſhould be ready to aſſemble and march upon the firſt notice; none ſhould paſs the gates, but who have ſome viſible buſineſs, or can give a good account of themſelves; the centinels ſhould not let any coaches, waggons, carts, *&c.* enter too cloſe behind one another, and when they are loaded with hay, ſtraw, or any other thing, wherein people may be concealed, they ſhould be well examined before they paſs; and never permit any carriage to ſtop upon a draw-bridge, on any pretence whatſoever, to prevent their being drawn up, if occaſion requires.

On

On holidays, feftivals, or rejoicing-days, the Governor fhould take particular care, to fee the guards kept in the ftricteft manner, and military difcipline obferved with the utmoft rigour; no great affemblies fuffered after dark.

The clergy are the moft apt to betray Places, as experience has fhewn, the Governor fhould frequently examine churches and religious houfes, to prevent conceal'd perfons, or whether they have any under-ground paffages leading out of the Town, as fometimes have been. Had the Governor of *Cremona* taken thefe precautions, he had not been furprized in bed as he was by Prince *Eugene,* who held private correfpondence with a prieft, that treacheroufly concealed a body of men in a chapel, and others that were let in to the Town, thro' an old aqueduct.

When the enemy knows that a Governor takes fuch and other precautions, they will hardy attempt the furprizal of a Place, and fhould they be adventurous enough to undertake it, it muft certainly turn to their difadvantage.

Of CAPITULATIONS.

THE Capitulation being the laft action, both in the Attac and Defence, this is the moft proper place to treat of it.

When a Governor, who defends a Place, fees himfelf reduced to the laft extremity, or is ordered by his Prince to furrender, to get better conditions from the enemy, and a more advantagious compofition, both for the inhabitants and garrifon, he orders to beat the *Chamade*; for which one or more drummers are to beat their drums on the rampart next to the Attac

to

to give notice to the Beſiegers, that the Governor has ſome propoſals to make; there are likewiſe put up, one or more white colours, upon the rampart, for the ſame purpoſe, and one of them remains either on the Breach or rampart, during all the time of negotiation. The ſame is done for asking a ſuſpenſion of arms, to bury the dead, and carry off the wounded, after a violent Attac.

The *Chamade* being beat, the fire ceaſes on both ſides, and the Governor ſends ſome officers of diſtinction to the Commander in chief of the Beſiegers, who deliver to him the conditions on which the Governor propoſes to ſurrender the Town. But as a ſecurity for the officers ſent from the garriſon, the Beſiegers ſend a like number into the Town. When the Governor's propoſals are not ſatisfactory to the General of the Beſiegers, he preſcribes the conditions on which the Town is to ſurrender; he commonly threatens the Governor to allow him no conditions at all, in caſe he refuſes thoſe propoſed, within a certain time, or when ſuch or ſuch a work is finiſhed. If the Beſieged find the conditions of the Beſiegers too hard, the officers return to their homes, and the drums are beat upon the rampart, to make every body retire before hoſtility begins, which is done in a very ſhort time after. It is to be obſerved, that during the ſuſpenſion of arms, no work ſhould be done on either ſide, either to ſecure the Beſieged or the Beſiegers; nothing ſhould be undertaken during the negotiation, yet it is very neceſſary to be upon the watch at that time as much as at any other, for fear of being ſurprized by ſtratagem, which is now looked upon as lawful.

But let us ſuppoſe, that the terms of capitulation are agreed upon; the Governor ſends two or three

of

of his principal officers into the camp, and the General ſends the ſame number, and of the ſame rank into the Town, as a ſecurity for accompliſhing the Capitulation. When the Beſieged have performed every thing according to agreement, their hoſtages are ſent back; and when the Beſiegers have performed every thing agreeable to the agreement, their hoſtages return.

The conditions of the Beſieged may be of various kinds, according to the different circumſtances or ſituations in which they are; the moſt common ones are,

1. That the Garriſon ſhall march out thro' the Breach, with their arms, baggage, horſes, drums beating, matches lighted on both ends, flying colours, a certain number of cannons and mortars, with their appurtenances, ſpare carriages, ammunition for a certain number of charges, to be conducted in ſafety to the Town agreed on, and which is uſually the next belonging to the Beſieged: it muſt be obſerved to inſert, by the ſhorteſt road, or the road is ſpecify'd in words, which the garriſon are to march. When the garriſon have ſeveral days to march before they can reach the Town agreed on, it is required that the troops ſhould be provided with proviſion and lodgments during that time.

2. One of the gates ſhall be delivered up to the Beſiegers, either the ſame evening, or at a certain hour next day, and the garriſon ſhall march out in a day or two after, according to the agreement made between both parties.

3. The Beſiegers ſhall furniſh a certain number of cover'd waggons, that is, which are not to be ſearched, beſides others to carry the wounded and ſick, which are in a condition to be tranſported; and,

and, in general, all the carriages neceſſary to carry the garriſon's baggage, and the artillery allowed by the capitulation.

4. That the ſick and wounded, which cannot be carry'd off, and are obliged to remain in the Place, ſhall have free liberty to go away with every thing that belongs to them, when they are in a condition to do it; and they ſhall be furniſhed, in the mean time, with lodgings and proviſion *gratis*, or otherwiſe.

5. There ſhall be no indemnification required from the Beſieged, for horſes taken from the inhabitants, or for houſes burned or deſtroyed during the ſiege.

6. The Governor, the officers under him, and thoſe belonging to the Garriſon, the Garriſon, and, in general, every one in the king's ſervice, ſhall freely go out of the Place, without repriſals of any kind.

7. If thoſe who take the Town are of a different religion from that of the inhabitants, it muſt be inſerted in the capitulation, that the inhabitants ſhall exerciſe their religion without any moleſtation.

8. That the inhabitants, and thoſe depending on the Place, ſhall be maintained in all their rights, privileges, and prerogatives.

9. It ſhall be at their choice, who have a mind to leave the Place, to go where they pleaſe with all their effects. It ſhould be ſtipulated, that the inhabitants who have ſhewn any partiality to the Garriſon, ſhall not be moleſted on that account.

10. It is alſo mentioned in the capitulation, that all the powder and ammunition remaining ſhall be delivered to the Beſiegers, and the loaded mines are likewiſe ſhewn.

11. That

11. That all the priſoners made on both ſides during the Siege ſhall be releaſed.

A Garriſon ſhould have proviſion and ammunition, at leaſt for three days, to be entitled to a compoſition, without which they are to be made priſoners of war: but if the Beſiegers have not inquired into it before the capitulation is ſigned, it would then be an injuſtice to make them priſoners of war.

When the Beſiegers will agree to no other compoſition than that the Garriſon ſhall be made priſoners of war, and the garriſon is not in a condition to hold out any longer, it is generally endeavoured to make the conditions as little onerous as poſſible.

1. That the Governor, and the principal officers, ſhall keep their ſwords, piſtols, baggage, *&c.*

2. That the ſubalterns, under the captains, ſhall keep their ſwords, with their baggage.

3. That the common men ſhall not be rifled, or diſperſed from their regiments.

4. That the Garriſon ſhall be conducted to a certain Place, by the ſhorteſt road, where they are to remain priſoners of war.

5. That the principal officers ſhall have leave for two or three days to go where they pleaſe, to ſettle their affairs.

6. When the Garriſon quits the Place, it ſhall not be permitted to decoy the ſoldiers to deſert from their regiments.

When the capitulation is ſettled, an officer of Artillery from the Beſiegers comes into the Place, with an officer of Artillery from the Garriſon, who take an inventory of all the Artillery and ammunition remaining in the Place; a commiſſary of proviſion provifion enters likewife, to take an account of the provifions.

When it is found neceffary to furrender, and that there are confiderable magazines ftored with ammunition and provifion, moft of it fhould be deftroyed before any mention is made of capitulating, that there may remain no more than what is neceffary, that the enemy may reap no benefit by them. If this fhould be done after the capitulation is mentioned, the Befiegers may infift on not fpoiling them; but what is already done cannot be avoided.

So foon as the befieged have delivered a gate of the Place to the Befiegers, the firft regiment of the army enters, and mounts guard.

The day on which the Garrifon is to leave the Place being come, the Befiegers army is put under arms, and ranged into two files, between which the Garrifon paffes. The time of marching being come, the General, and the principal officers, head the two files, to fee the Garrifon defile before them.

The Governor marches at the head, followed by the principal officers, who make the Garrifon march in the beft order poffible. The eldeft regiments march commonly the firft and laft, and the reft in the center with the baggage. When there are any horfe, they are alfo divided into three bodies, to march at the head, center, and in the rear. Small detachments of horfe and foot are made, to march at the fides of the baggage, to take care of its not being rifled.

The Artillery allowed by the capitulation march after the firft battalion.

When the Garrifon is arrived at the Place agreed on, the Governor remits the hoftages of the Befiegers ſiegers to the eſcorte; and when the eſcorte is arrived at the army, the hoſtages which the Beſieged have left for the ſecurity of the eſcorte, carriages, and other things allowed by the army for eſcorting the Garriſon, are releaſed.

When the Garriſon is made priſoners of war, they are likewiſe eſcorted to the Place agreed on.

Every thing agreed on ſhould be looked upon as ſacred and inviolable, and every word ſhould be underſtood in its genuine ſenſe, without any forced conſtruction; yet as this is not always the caſe, the Governor ſhould be very cautious not to have any word inſerted, but what is clear and plain, without admitting of any other ſenſe than that for which it is uſed. There are many examples which prove the neceſſity of this precaution.

In the capitulation of a Garriſon, where there is a Citadel, in which the Garriſon retires, there are ſome particular conditions to be requeſted, ſuch as follow.

That the Citadel ſhall not be attacked at that ſide next to the Place; that the ſick and wounded, which cannot be tranſported, ſhall remain in the Place, and in the lodgings where they are; and after being cured, they ſhall be provided with carriages and paſſports, to retire to the place agreed on in the capitulation. No body ſhould be let into the Citadel, but thoſe who may be uſeful in its Defence; the reſt, who are uſeleſs, ſhould be no means be ſuffered to enter. It muſt be mentioned in the capitulation, that theſe people ſhall be conducted to a neighbouring place belonging to their Sovereign, which is to be named. It ſhould alſo be agreed on, to have a certain time allowed for the Garriſon to march into the Citadel; and it ſhould be be abſolutely prohibited that the Beſiegers ſhall make no works whatever, for carrying on the Approaches towards the reduction of the Citadel, during the time preſcribed.

A maritime Town requires likewiſe ſome particular conditions relating to the Ships which may be in the harbour. It ſhould be agreed, that they ſhall leave the Harbour the ſame day that the Garriſon leaves the Place, or when the weather permits, to ſail to the port agreed on. They keep all their artillery, ammunition and proviſion, &c. If the bad weather ſhould oblige them to enter any harbour belonging to the Beſiegers, it ſhould be mentioned in the capitulation, that they ſhall be received, and that they ſhall be furniſhed with neceſſaries to continue their voyage; they ſhould be provided with paſſports, and, in ſhort, all the ſecurity poſſible, not to be inſulted by the enemy's ſhips, till they are arrived at the port ſpecified.

Many other things may be ſaid, with regard to the ſubject treated on in this work; it would require a larger volume, to enter into every particular; all that has been ſaid ſhould be regarded only as a ſummary account of the principal attentions which it requires, and what is moſt generally obſerved.

Beſides, as a late Author with reaſon obſerves, Places have different Defences, according to their ſituations, and as they are defended with more or leſs forces, the experience and courage of a Governor ſhould ſuggeſt to him the beſt Defence, to furniſh him with reſources to repair any accidents that may happen, and to make the beſt advantage of the Beſiegers miſtakes and negligence.

It

It is not ſufficient to have courage enough to defend the Place well, it requires beſides great ſagacity and knowledge, not only in the art of War, but likewiſe in Fortification.

The Defence is attended with many more difficulties than the Attac, and it may conſequently do more honour to a General, who diſtinguiſhes himſelf in it. The ſuperiority over an enemy in an Attac, the conveniencies there is in receiving freſh forces and ammunition, whenever it is wanted, and all other neceſſaries, which may be had from the neighbouring country; all this may ſerve to repair any accidents that may happen during the Siege. It is not ſo in the Defence, no faults are committed unpuniſhed in the face of an underſtanding enemy. The attention muſt be equally over the ſoldiers and inhabitants; keep a ſtrict watch within and without; not to expoſe the troops without great neceſſity, but upon ſuch occaſions as are viſibly uſeful. In ſhort, the Governor muſt create a reſpect from the enemy, by his conduct and ſagacity; never give up the leaſt part of the works, till after having exhauſted all poſſible ſtratagems in the maintaining them. All this requires a great capacity.

A TREA-

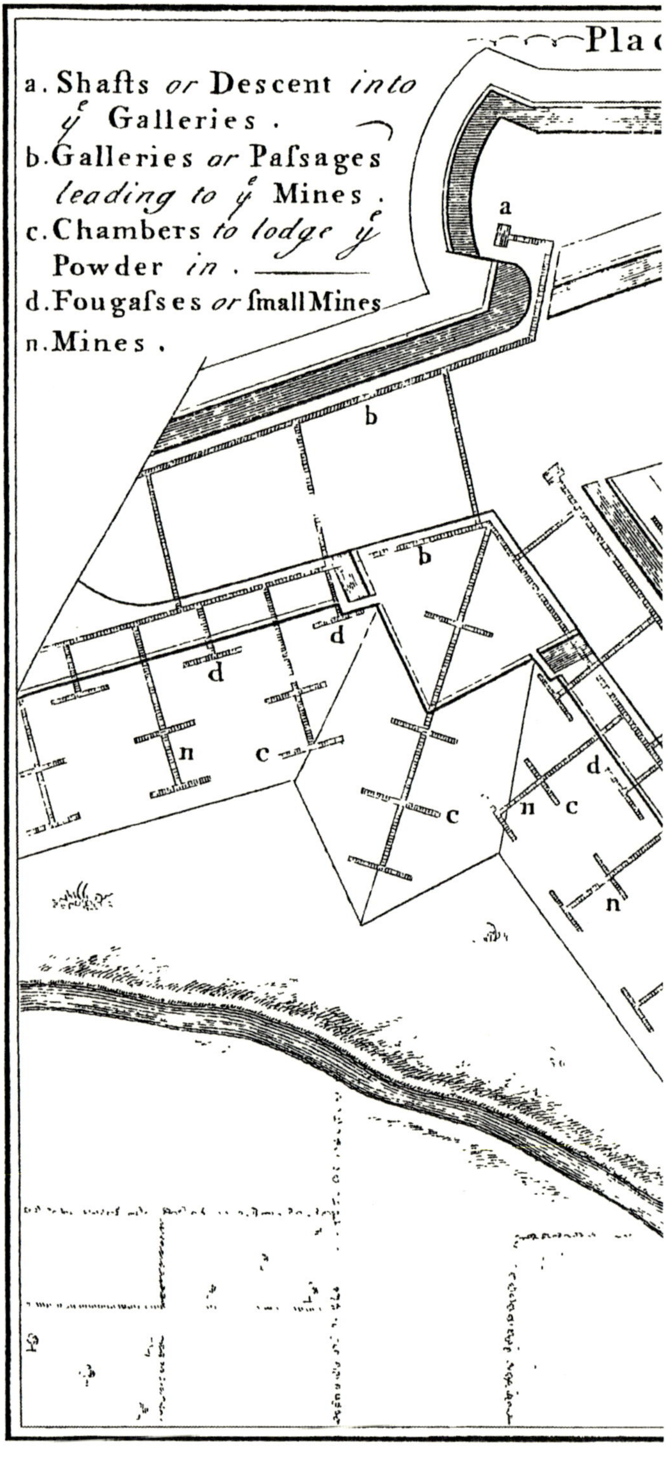

Pla
a. Shafts or Descent into ye Galleries.
b. Galleries or Passages leading to ye Mines.
c. Chambers to lodge ye Powder in.
d. Fougasses or small Mines
n. Mines.
a
b
b
d
d
n
c
c
d
n
c
n

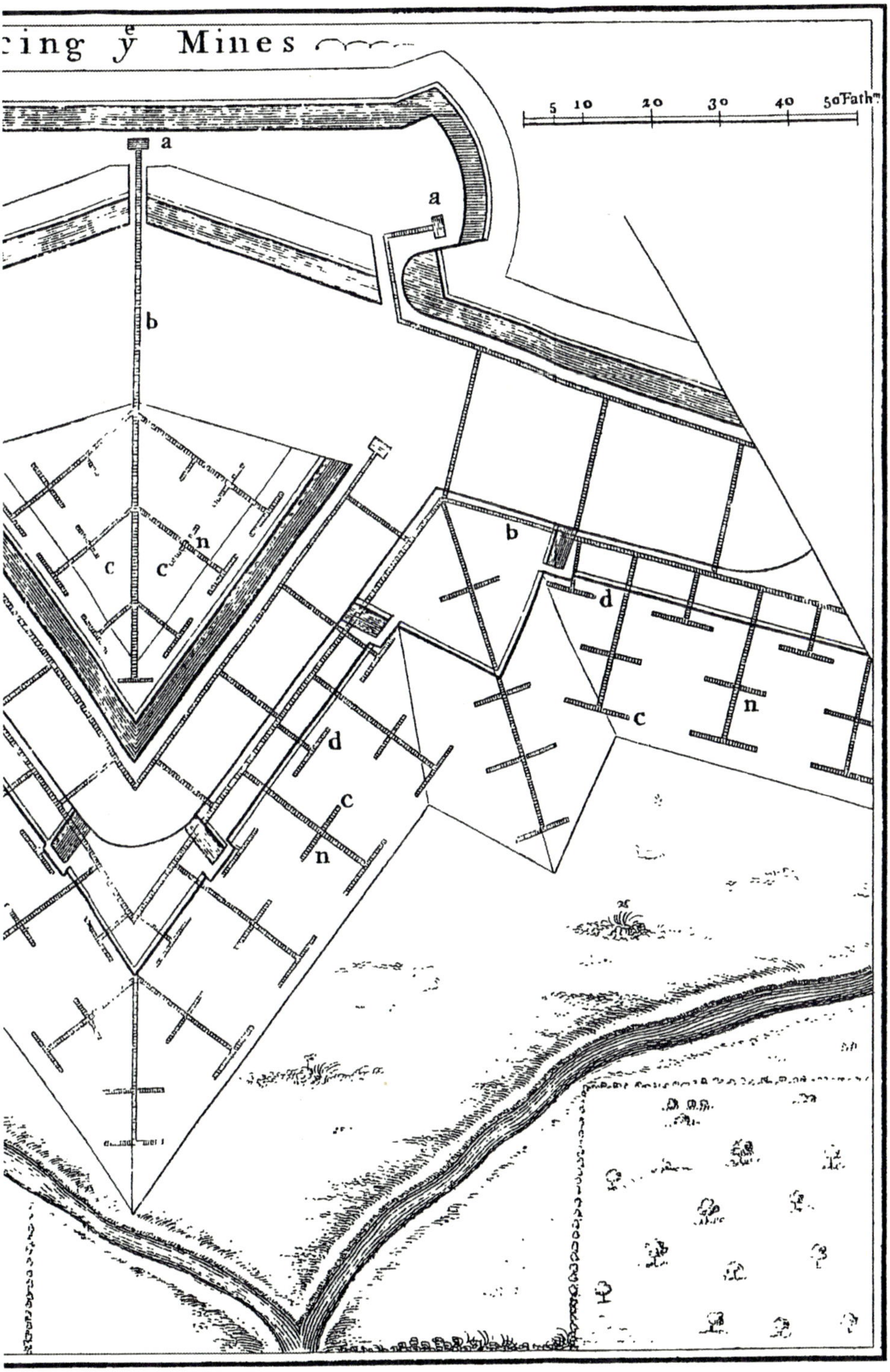
cing y^e Mines
5 10 20 30 40 50 Fath^m
a
a
b
b
c
c
c
c
c
n
n
n
d
d

p. 204.

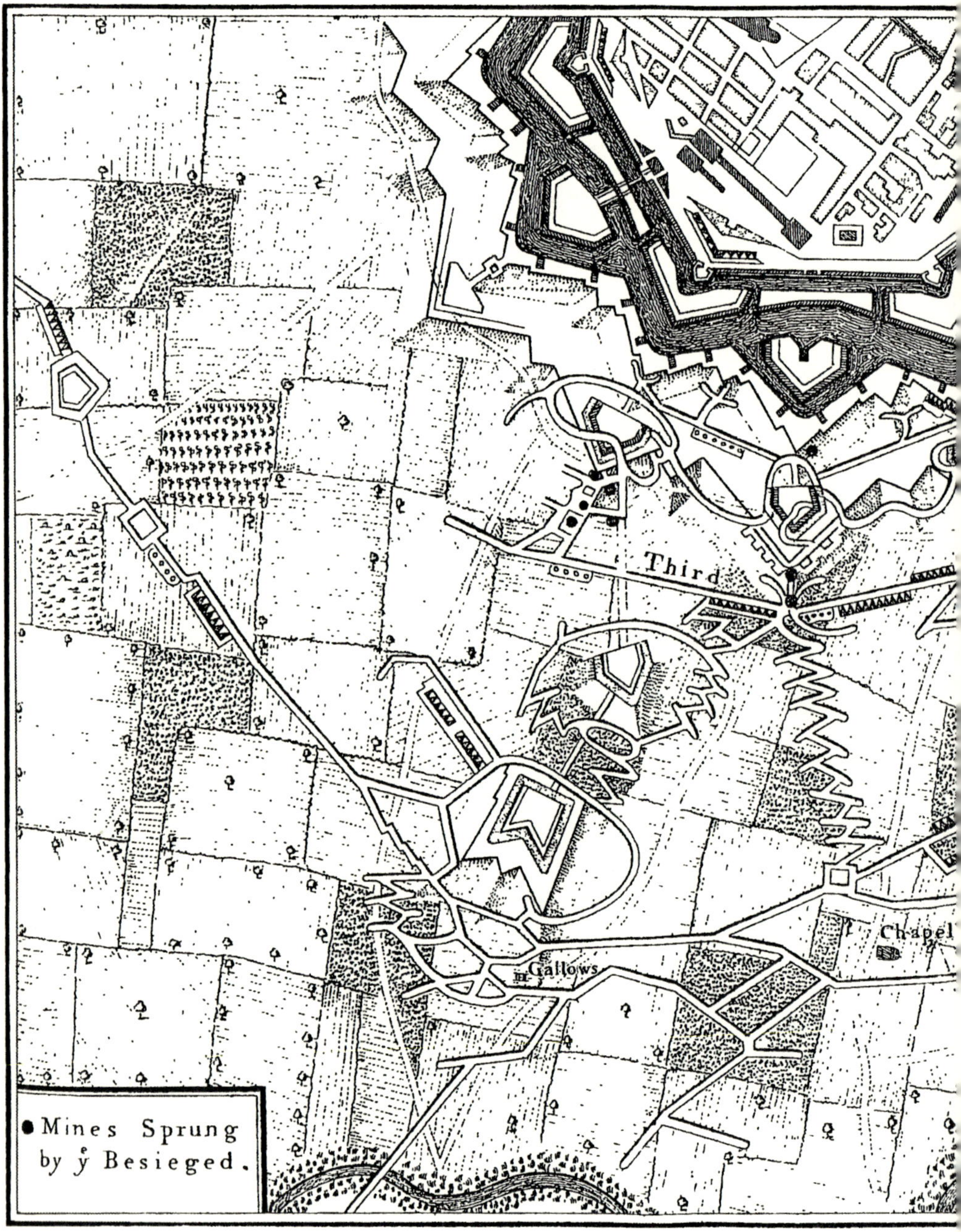

XXIII.

ATTACKS of
LANDAU in 1713.
Queiche R.
30 60 90 180 Fathom
Parallel
Parallel
Parallel
Second
First

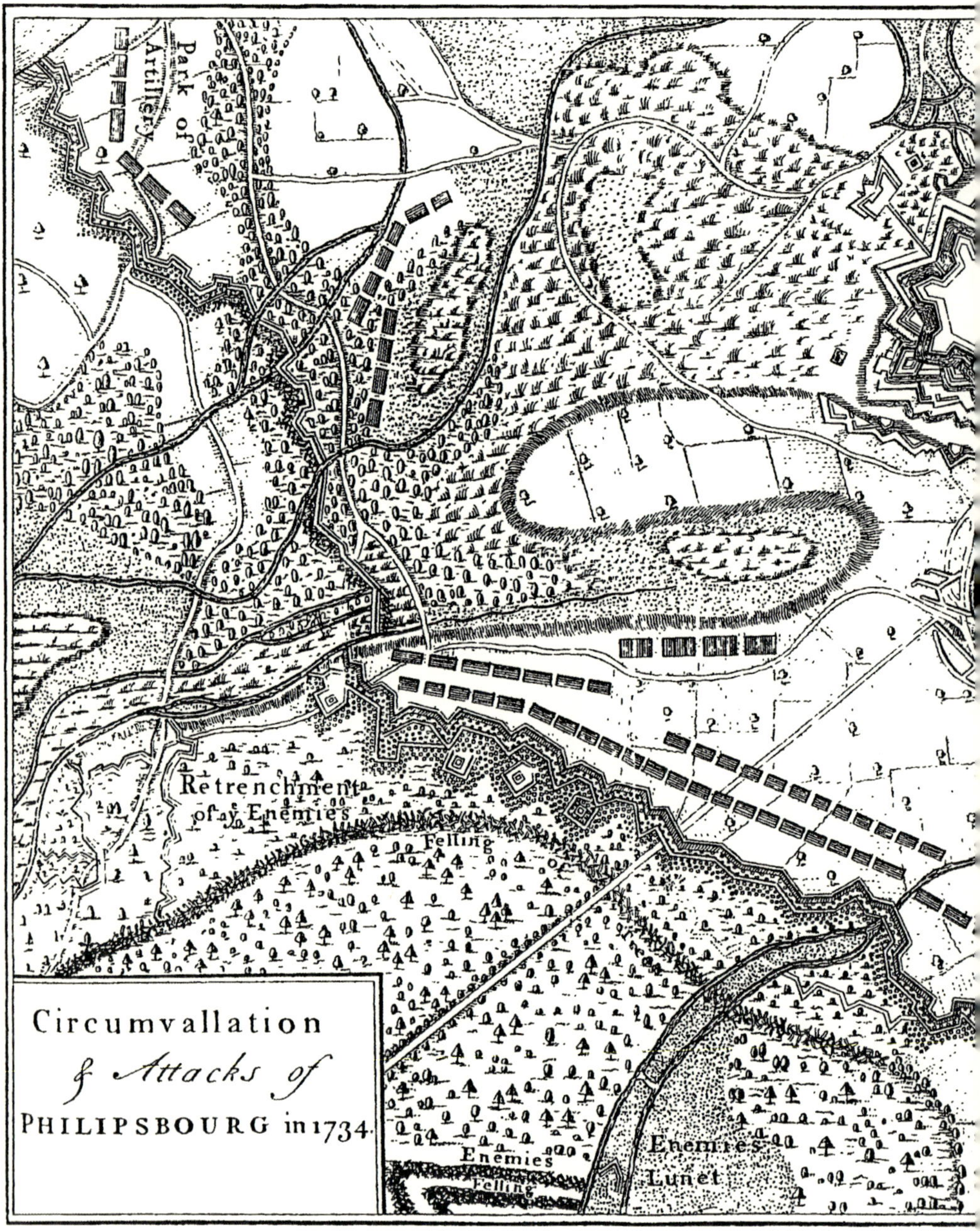

Circumvallation & Attacks of PHILIPSBOURG in 1734.

RHINE R
Green Redout
Capuchin Redout
Oberhausen
Genl Quarter
Great Guard of Dragoons
Wagheusel
Great Guard of Dragoons
50 100 200 300 400 500 600 Fath.

A

TREATISE

OF THE

ATTAC and DEFENCE

OF

FORTIFIED PLACES.

PART III.

Of MINES.

THE Art of Mining is become one of the moſt eſſential parts of the Attac and Defence of Places: ſo much Artillery is uſed, that nothing above ground can withſtand its effect; for the moſt ſubſtantial ramparts and parapets can reſiſt but a very ſhort time: the out-works, tho' numerous, ſerve no other purpoſe than to retard, for ſome time, the ſurrender of the Place; and that advantage is paid dear

dear for, by the great number of troops wanted to defend them, and the immenſe quantity of ſtores and ammunition expended.

The object of a Fortification being, that a ſmall number of men ſhall be enabled to defend themſelves advantageouſly againſt a conſiderable army; but if a Place requires nearly ſo many men to defend it as to attac it, a great expence is thrown away to no purpoſe, which will ever be the caſe if it is not countermined.

Whereas a moderate garriſon, ſufficient to guard the ſeveral works, with a company of miners of 60 men, will be able to deſtroy the Beſiegers and their works, ſo ſoon as they come near or upon the glacis; this they may do at pleaſure, provided proper galleries have been made, and that the miners know properly to diſpoſe of the ground. This the Beſiegers can only prevent, by carrying on their approaches under ground as well as above, which requires infinite time and labour. They are then unequal to the Beſieged; they having their principal galleries prepared before, whereby they are enabled to deſtroy them and their works, whenever they attempt to come near. Therefore the true art of fortifying depends upon that of mining, as well as the Attac and Defence of Places.

Having explained, in the two preceding parts of this work, every thing relative to all the different works made above ground, by the Beſiegers and Beſieged; I ſhall endeavour here to treat as full of thoſe made under ground. I have ſhewn the uſe of Mines, and the ſeveral ſtratagems of the miners, it remains to explain the manner of carrying on the Galleries according to a given direction, that of loading them with a proper charge, in reſpect to their

their depth and different tenacities or textures of the ſoil they are made in, together with the diſpoſitions of the chambers, ſo as to leave nothing unexplained, and to render the art compleat.

Before I enter upon theſe particulars, it will not be improper to ſay ſomething relating to the riſe and progreſs of this art, as likewiſe to the various opinions of the cauſes which contribute to their effect, and of the form of the pit or hole they make when ſprung. We find in hiſtory that Mines were made long before the invention of Gun-powder; for the antients made galleries or under-ground paſſages much in the ſame manner as the moderns, from without, under the walls of places, which they cut off from the foundation, and ſupported them with ſtrong props, then they filled the intervals with all manner of combuſtibles, which being ſet on fire burnt their props, and the wall being no longer ſupported fell, whereby a Breach was made.

The Beſieged alſo made uſe of under-ground paſſages from the town under the Beſieger's machines, whereby they batter'd the walls at that time to deſtroy them; which proves neceſſity to be the inventor of Mines, as of many other uſeful arts.

Since the invention of Gun-powder, the art of mining has receiv'd great improvements, the work has been much ſhorten'd, and their effects are become more ſudden and terrible; nothing can reſiſt the force of the powder, the ſtrongeſt rocks are not proof againſt its irreſiſtible force, witneſs the caſtle of *Alicant*, which was deſtroyed by the *French*; and many ſimilar effects are well known thereby to have been produced.

The firſt Mines, ſince the invention of Gun-powder, were in 1487, by the *Genoeſe*, in the

Attac

Attac of *Serezanella*, a town belonging to *Florence*; but theſe failing, they were ſome time neglected, till *Peter Navarro*, according to Mr. *le Blond*, being then Engineer to the *Genoeſe*, and after to the *Spaniards* in 1503, againſt the *French*, at the ſiege of the caſtle called *del Ovo*, or Egg, at *Naples*, made a Mine under the wall, and the Governor was ſummoned to ſurrender, which not being complied with, he blew up the wall, and the caſtle was taken by ſtorm.

Mr. *Valliere* relates the ſame ſtory, but differs in the name of the engineer; for he ſays, it was one *Francis George*, an *Italian*, who ſerving at *Naples* in quality of architect, propoſed to *Peter Navarro*, the *Spaniſh* General, to take this caſtle by Mines.

Since then Mines were made in the Attac and Defence of moſt Places: but as the miners had only practical knowledge, this art received but ſmall improvement, till towards the end of the laſt century; the continual wars at that time in different parts of *Europe*, in which the *French* were generally concerned, induced their miners to make tables for the charges of Mines, according to their depth under-ground. But as they had not ſufficient skill to eſtimate the quantity of earth to be blown up, theſe tables were very imperfect.

The great engineer M. *Vauban*, was the firſt that wrote upon Mines, but that art, being then in its infancy, he did not make that improvement which might have been expected from his great genius in the art of war: yet the world is indebted to him, for endeavouring to reduce this art into a ſcience, and to induce others to make farther inquiries: after him ſucceeded the celebrated *Valliere*, lieutenant

and

and inſpector general of the artillery, who being one of the greateſt maſters in the art of mining, and joining the theory to a long ſeries of practice, made a great ſtep towards rendering this art compleat. He was ſenſible, that the pit or hole made in the earth when the Mine is ſprung, was not an inverted cone, nor a fruſtrum of a cone, as ſuppoſed by *Vauban* and others before him: by meaſuring ſeveral excavations, he found that it was a Paraboloid nearly; and his tables were computed accordingly to that figure: he alſo wrote a memoir concerning the Mines made by the Beſieged under the glacis and covert-way, with ſo much knowledge and judgment, that it will remain an eternal monument to his memory. Yet, notwithſtanding theſe ſeveral improvements, in the art of mining, it was far from being compleat, as ſome unthinking authors have lately advanced, there remaining ſtill one great prejudice to ſurmount, ſo ſtrongly imprinted in the minds of the practical miners all over *Europe*, that it can ſcarcely be eradicated by the moſt glaring evidence. Which is, that the diameter of the pit or hole made by the Mine, was always twice the line of leaſt reſiſtance, or that line drawn from the center of the powder, perpendicular to the neareſt ſurface, and that the before-mentioned diameter, could never exceed twice this line.

The chevalier *Belidor*, formerly profeſſor of artillery at *La Fere*, undertook to remove this prejudice; and tho' he proved by many experiments, that the diameter of the hole made by a Mine may be increaſed to any length in regard to the depth of the Mine, yet very few miners are convinced of it at this time: it is true that the memoirs he wrote on the experiments made at *La Fere*, had never

been

been publiſhed; yet moſt authors have mentioned ſome parts of them in their works: but as they knew nothing of Mines, having never ſeen any, they did no more than copy other writers; they ſpeak of this improvement ſo as to leave the reader in doubt, whether it deſerves to be taken notice of or not. At the ſame time they give a copy of a diſſertation on Mines, publiſhed ſome years ſince in his courſe of mathematics, endeavouring to perſuade the reader, that it is the moſt compleat treatiſe of Mines ever wrote: tho' he acknowledges not to have had ſufficient experience when he wrote it, having ſince found many things inconſiſtent with truth, begs the reader not to lay any ſtreſs upon it.

Belidor was preſent when *Valliere* tried the experiments for diſcovering the true figure of Mines; and made ſome himſelf upon the ſame account, yet in his manuſcript memoirs, he adopts another method for finding the charges of Mines; whether he concluded that this figure would not anſwer all the different problems in the art of mining, or that it was unneceſſary to conſider the figure at all, is a queſtion I am unable to decide.

The ſame gentleman made ſome experiments ſince by orders from court, concerning a new method for making the Approaches at once, from the extremity of the glacis to the covert-way by Mines, thereby not only to diminiſh the time and expences of a Siege, but alſo to ſave many lives. This and much more may be done by the Beſieged, ſuppoſing the nature of the ground will admit of digging twenty feet deep without meeting water. The diſpoſition of the Counter-mines mentioned by *Valliere*, have been eſteemed the beſt that are poſſible to be made, yet I am certain, that much more

more may be done with very little labour and leſs expence; it appears to me that a common fortification with baſtions, ravelins and a covert-way, may be made ſo ſtrong by means of Counter-mines, as to render it almoſt impoſſible to take it, or at leaſt that if the Beſiegers be obſtinately bent to take the Place, their loſs would be more conſiderable than their gain.

Thoſe who imagine that the art of mining is brought to the utmoſt perfection, are greatly miſtaken; all that has hitherto been done, depends upon ſuch principles as are derived from practice and common experience, which alone by a nearer inſpection will be found inſufficient; a theory deduced from the nature and effects of gun-powder, and built upon ſuch mechanical principles as are taught by the mathematics, well ſupported and confirmed by proper experiments, is the only way to render that art compleat. As no author has connected the theory and practice together, except *Belidor*, and tho' he has freed the art from many abſurd notions of the practical miners, yet he could not intirely avoid adopting ſome erroneous opinions received from his predeceſſors.

In the following work, will be conſidered, the two different methods, whereby the charges of Mines are computed, that is, when the pit or hole is taken for a paraboloid, or when the ſphere of exploſion is uſed; and by comparing both with the few experiments that may be depended on, the reader will be able to judge, which ſhould be preferred: I ſhall likewiſe ſubjoin M. *de Valliere*'s tables, and with regard to Mr. *Belidor*'s laſt experiments, and ſome particular things, they are

things,

things, which for particular reaſons, I hope to be excuſed from mentioning.

DEFINITIONS.

BY a *Mine* is underſtood a paſſage carried under-ground to the Place that is to be blown up, where a lodgment is contrived for the powder, neceſſary to produce the propoſed effect.

The paſſage leading to the powder, is called the *Gallery*.

The place where the powder is lodged, the *Chamber*.

The line, drawn from the center of the ſpace, containing the powder, perpendicular to the neareſt ſurface, is called the *Line of leaſt reſiſtance*.

The pit or hole made by a Mine when ſprung, is called the *Excavation*.

The fire is given to the Mines by a pipe or hoſe, made of coarſe cloth, whoſe diameter is about an inch and a half, called *Sauciſſon*, extending from the chamber to the entrance of the gallery, to the end of which is fixed a match, that the miner who ſets fire to it, may have time to retire before it reaches the chamber.

To prevent the powder from contracting any dampneſs, the ſauciſſon is laid in a ſmall trough, called *Auget*, with ſtraw in it, and round the ſauciſſon, with a wooden cover nailed upon it.

The Mines made by the Beſiegers in the Attac of a Place, are ſimply called *Mines*, and thoſe made by the Beſieged *Counter-mines*. Both Mines and Counter-mines are made in the ſame manner, and for the like purpoſes, *viz*. to blow up their enemies and their works; only the principal galleries and Mines

mines of the Befieged, are ufually made before the town is befieged, and frequently at the fame time, that the fortification is built, to fave expences. The Befieged generally make a great many fmall Mines under the glacis, of about 6, 7 or 8 feet deep under-ground only, which are called *Fougaffes.* They make likewife another fort, called *Coffers* or *Caiffons,* a kind of wooden boxes 3 or 4 feet long, and a foot or 18 inches wide, which they bury 4, 5 or 6 feet under the glacis, and about 4 yards diftant from each other.

The whole art of mining confifts in the four following particulars, *viz.*

1. In making the galleries and chambers, and fecuring the earth from falling in, by means of wooden frames or props.

2. In the loading and ftopping the Mines, fo as to be ready for fpringing when required.

3. In knowing the proper charges for Mines placed in different foils, or at different depths under-ground, that the effects may anfwer the intentions for which they were made.

4. Laftly, in carrying on the galleries according to a given direction, to a given length, fo as to place the chamber directly under the object to be blown up, at a certain depth under-ground.

Thefe particulars fhall be treated of as amply as the fubject requires, and each feparately according to their order here mentioned.

Of making GALLERIES *and* CHAMBERS.

WE fhall omit the manner of fixing the miner to the wall, to make a Breach, and how to fink fhafts in or near the third Parallel, thefe having been explained in the Attac of Places.

The

The Galleries made within the fortification, before the Place is attacked, and from which ſeveral branches are carried to different places, are generally 4 feet wide and 5 high, and the earth ſupported from falling in by arches and walls, as they are to ſerve a conſiderable time; if they were ſupported with props or frames of wood, they would continually want repairing, which would at length coſt more than ſtone walls; but when Mines are made to be uſed in a ſhort time, then the Galleries are only about three feet wide and 5 high, and the earth ſupported with wooden frames or props.

The entrance of the Mine being made, and the direction of the Gallery known, the firſt miner works upon his knees, whilſt another is behind him, to gather up the earth into a basket or wheelbarrow, which he carries or draws out, and throws near the entrance.

If the entrance is a ſhaft or pit ſunk pretty deep, there muſt be another miner above-ground to let down a ſmall rope, with an iron hook faſtened to the end, into which the miner below hands the basket, then fetches another while the firſt is emptied.

When the ſhaft is very deep, and the Gallery pretty long, a capſtan is fixed over the ſhaft for greater expedition, and a couple of baskets are faſtened to a rope, wound round the axis of the capſtan; that whilſt one draws up the other goes down, as water is drawn out of a well with buckets.

The firſt miner being tired, the next to him reſumes his place, and he goes laſt, they ſucceed one another thus till all are fatigued, then they are

are relieved by an equal number, which is generally in two hours time.

When the firſt miner digs up more earth than one man can carry off, another ſtands behind him at ſome diſtance, who takes the basket from the firſt, and carries it to the entrance; and as the length of the Gallery increaſes, more men are placed in it to carry out the earth; it muſt be obſerved, that they are always placed at equal diſtances from each other, to reach the baskets from hand to hand, and prevent confuſion.

The Gallery being carried on ſome diſtance, carpenters follow the miners to prop the earth above, if it is not ſtiff clay or loom, and to prevent its falling in. This is done by placing a piece of wood a-croſs the Gallery over-head, and putting a prop or upright at each end, faſtened into the ground, then the earth is removed above this piece, ſo as to ſlip a piece of deal board over it, which being afterwards ſupported in the ſame manner at the other end, will keep up the earth.

Sometimes wooden frames are made, of the ſize of the Gallery, and placed at proper diſtances from each other, and then boards are ſlipped over them; and to make the frames or uprights tight and firm, wedges are drove underneath, otherwiſe the ſhaking of the ground ariſing from the firing of the guns and mortars, or from the ſpringing ſome continguous Mines, may occaſion the earth to fall in, and endanger the lives of the miners.

Whilſt the work is carrying on, the miners uſe a maſon's level, to level the bottom of the Gallery, to keep to the ſame depth; and when they come to any turnings, they make uſe of a ſquare to make them

them at right angles; which they likewiſe examine with their needle and compaſs.

The Gallery being carried on to the place where the powder is to be lodged, the miners make the Chamber, which is generally of a cubical form, large enough to hold the wooden box, which contains the powder neceſſary for the charge, the box is lined with ſtraw and ſand-bags, to prevent the powder from contracting any dampneſs.

The Chamber is ſunk ſomething lower than the Gallery; if there is room enough to place the box, load the Mine, and buttreſs the upper part well, it is ſufficient. It muſt be obſerved, that when the Beſieged can raiſe the water in the ditch, and the Gallery can be incommoded thereby, inſtead of making the Chamber lower than the Gallery, it muſt be made higher, otherwiſe they will certainly let in the water and ſpoil the Mine.

When the Gallery is very long and narrow, it often happens that the air is ſo ſtagnated, that the candle will not burn, and the miners can hardly breathe, which obliges them to come out often in the open air. Many inventions have been uſed to prevent this inconveniency, ſome make uſe of a large pair of bellows, to the mouth of which they fix a leather pipe or hoſe, of about three inches in diameter, which goes from the entrance of the Gallery to the farther end, where the miners work; or elſe they ſuſpend a Sac made in the form of a funnel, of about two feet diameter above, and three inches below hooped within, at every foot diſtance, to keep the cloth ſtretched, and fix a leather pipe to it as before: but this method will, in my opinion, not anſwer; for the difference between the rarefaction of the air above

and

and below, is not ſufficient to cauſe a ſenſible circulation. The beſt way, is to bore holes upwards to the ſurface of the earth in a conical form, at every 40 or 50 yards diſtance: this may partly be done with an augur of ſeveral pieces and then widened below: but this cannot always be done, eſpecially when the gallery is made under a high rampart or buildings, and where the enemy may perceive it, who would not fail to throw ſome ſtinking compoſition into the gallery to ſmother the miners. I have been informed, that a wooden pipe laid from the entrance to the further end will cauſe a ſufficient circulation of air that the miners may work without inconveniency: ſome affirm, that a fire of charcoal placed at the entrance will anſwer; this I much doubt, unleſs a pipe is fixed to it. It ſhould be obſerved, that when a leather pipe is uſed, it muſt be hooped within to keep it open, that the air may have a free paſſage, otherwiſe it will never anſwer the purpoſe.

Of loading and ſtopping the Mines.

It would appear more regular to treat of the proper charges of Mines, before mentioning the manner of their loading; but as the computing the charge is a diſquiſition of an important nature, and chiefly depends on mathematical demonſtrations, I thought that by treating the practical parts immediately one after another, would not be improper.

The gallery and chamber being ready to be loaded, a ſtrong box of wood is made of the ſize and figure of the chamber, being about a third or fourth part bigger than is required for containing the

the neceſſary quantity of powder; againſt the ſides and bottom of this Box is put ſome ſtraw, and this ſtraw is covered over with empty ſand-bags, to prevent the powder from contracting any dampneſs; a hole is made in the ſide next to the gallery, near the bottom, for the ſauciſſon to paſs through, which is fixed to the middle of the bottom, by means of a wooden peg, to prevent its looſening from the powder; or that in caſe the enemy ſhould get to the entrance, they may not be able to tear it out. This done, the powder is brought in ſand-bags, and thrown looſe in the box, and covered alſo with ſtraw and ſand-bags; upon this is put the cover of the box, preſſed down very tight with ſtrong props; and to make them more ſecure, planks are alſo put above them, againſt the earth, and wedged in as faſt as poſſible.

This done, the vacant ſpaces between the props are filled up with ſtones and dung, ſerving for mortar, and rammed in the ſtrongeſt manner; for the leaſt neglect in this work will conſiderably alter the effect of the Mine.

After this the auget, or wooden trough, is laid from the chamber to the entrance to the gallery, with ſome ſtraw at the bottom; then the ſauciſſon is laid in it, with ſtraw over it, and ſo ſhut with a wooden cover nailed upon it. Special care muſt be taken in ſtopping up the gallery, not to preſs too hard upon the auget, for fear of ſpoiling the ſauciſſon, which may hinder the powder from taking fire, and ſo prevent the Mine from ſpringing.

Then the gallery is ſtopped up with ſtones, earth and dung, well rammed, five or ſix feet farther from the chamber than the length of the Line of leaſt reſiſtance.

But

But to prevent the Mine from burſting through the gallery, and to make it have its effect upwards, the gallery is made with one or two turnings, at right angles to each other, and ſtrongly ſecured with buttreſſes and planks, and the intervals rammed with ſtones and earth. It muſt be obſerved, that the diſtance of the gallery ſtopped up, is to be eſteemed in a right line from the chamber, and not according to its turnings.

Proper Quantities of POWDER *to charge* MINES.

THE eſtimation of the proper quantity of Powder, with which a Mine is to be loaded in any kind of ſoil, or at any depth under-ground, to produce the expected effect, is the moſt difficult part of the whole art of mining. The innumerable variety of ſoils which are to be met with, the different texture or tenacities of their parts, even in ſoils, which in all appearance ſeem to be alike, together with the different ſtrata's of ſand, clay, ſtone, chalk, &c. leave hardly room or hope to arrive ever at any exactneſs in this art, without mentioning the inequalities of the ſtrength of gun-powder, which varies no leſs than the ſoils.

All that therefore can be expected is, to eſtabliſh a theory on ſuch experiments, as may be moſt depended upon, which, if not abſolutely true, may at leaſt be ſufficiently near, ſo as to anſwer in ſome degree of exactneſs in practice, and thereby avoid ſuch manifeſt errors, as many miners have been led into, by experiments made by perſons, whoſe great names cover their little knowledge in the mathematics, without which no true concluſion can be made from experiments.

The

The charging of Mines not only depends on the quantities of earth to be blown up, but likewiſe on the tenacities or difficulties to ſeparate their parts. Some inſiſt, that the tenacity is to be eſtimated by the ſurface of the excavation, as ſuppoſing the whole ſolid to be blown up entire, without any ſeparation of its parts, in which they are groſſly miſtaken. M. *Valliere* ſeems to be of this opinion in his writings, and yet does not make uſe of it in his tables. *Belidor* has inſiſted upon the ſame in the diſcourſe of Mines at the end of his mathematical courſe, but rejects it in his later writings, yet notwithſtanding, writers upon Mines inſert it in their books as a very good performance; for whoever has ſeen a Mine ſprung, muſt have eaſily perceived that all the earth blown up is ſeparated into ſmall parcels, except for ſome lumps, which are not divided, owing to their greater tenacity; and therefore the reſiſtance in the ſame ſoil is nearly proportional to the weights of the quantities of earth blown up.

The quantity of earth to be blown up depends on the figure of the excavation; that known, its ſolid content may ſoon be found by geometry; and by weighing exactly a cubic foot of that ſoil, it would be eaſily known what weight is to be raiſed; and by knowing what quantity of powder is required to raiſe a certain weight, the tenacity of the parts may likewiſe be found, by making a Mine ſo as to produce a good effect; and ſubtracting the quantity of powder, neceſſary to raiſe the weight of the ſolid from the charge of the Mine, the remainder would be the quantity neceſſary to overcome the tenacity.

Though it ſeems to be very poſſible to obſerve the true figure of the Excavation, by removing all the

the loofe earth, and meafuring the excavation exactly; and this being done in many foils and depths, one may expect to arrive at leaft pretty near, tho' not exactly, to the true figure.

Yet it is a matter of difpute what that Figure is; it was at firft imagined to be an inverted Cone, as A C B, whofe vertex being in the center of the chamber, and the radius of its bafe A D, to be equal to its Axis C D; but this being found to make the charge too fmall, they fuppofed it to be a fruftrum of a cone, fuch as AEFB, whofe leffer bafe E F, is equal to the line C D, of leaft refiftance, and the greater A B, equal to twice that line: and it is upon this fuppofition that the miners have computed their tables of the quantities of powder neceffary to charge Mines with, which are placed at different depths.

Fig. I.

Fig. II.

Mr. *Maigrigny*, an engineer of great experience, was the firft I know, who made fome experiments near *Tournay*, under M. *Vauban*; where he pretends to have found, that when a Mine was overloaded, inftead of producing a great effect, as might be expected, it would only make a hole like a pit; but that a certain quantity of powder would make the excavation a fruftrum of a cone, fuch as mentioned before; and that the radius of the greater bafe could never exceed the line of leaft refiftance, let the charge be what it will.

This fuppofition, not to fay fiction, is abfolutely falfe and contradictory to the firft principles of nature, that the effects are always proportionable to their adequate caufes; yet as abfurd as this notion is, it is ftill retained by practitioners and men of knowledge.

This

This is the caſe of unskilled practitioners, who receive the greateſt falſehoods for undeniable truths, without examining whether it be an impoſition or not; and reaſon, the greateſt gift beſtowed upon man, ſeems to be of no other uſe than to ſerve as a vain pretence to knowledge.

How far *Maigrigny* was qualified for experiments of this nature, will appear beſt by his manner of computing the quantities of earth to be blown up by Mines of different depths; he ſays, ſuppoſe it requires 100 lb. of powder to charge a Mine, whoſe line of leaſt reſiſtance is ten feet, then a Mine made in the ſame ſoil, whoſe line of leaſt reſiſtance is twenty feet, will require 200 lb. as having double the quantity of earth to raiſe; whereas theſe ſolids are ſuppoſed to be ſimilar, and therefore are as the cubes of their axis; that is, the ſecond would require 800 lb. to produce the like effect.

A proof this gentleman had not the leaſt knowledge in geometry, and therefore unable to make any right concluſion, or draw any conſequence from his experiments.

Belidor endeavoured in vain to ſhew the falſities of *Maigrigny*'s concluſions; he being oppoſed by all the practical miners, till at laſt it was reſolved by the chief commander of the Artillery at *Lafere*, to make ſeveral experiments, to know how far *Maigrigny*'s experiments might be depended upon; there were accordingly above 150 Mines made from 1725 to 1730, to many of which I was an eye-witneſs; the line of leaſt reſiſtance of the firſt was 10 feet, and the Mine was loaded with 1000 lb. of powder; which was above tripple the charge aſſigned by *Maigrigny*; and yet this

this Mine had ſuch a prodigious effect, as to blow moats or ſhoals of earth near a mile off.

As this Mine was only intended to prove that the greater the charge of a Mine was, the greater the effect it produced would be; it was no farther examined, it being thought ſufficient to convince all preſent, that *Maigrigny* had miſtaken the point in queſtion. After this, ſeven Mines were made, having all their line of leaſt reſiſtance, 10 feet; and loaded with the following quantities of powder, viz. the firſt, with 120 lb. ſecond, 160; third, 200; fourth, 240; fifth, 280; ſixth, 320; and the ſeventh, with 360: theſe Mines being ſprung one after another, and their excavations examined, the diameters of their baſes were found thus; that of the firſt, 22 $\frac{2}{3}$ feet; the ſecond, 26; the third, 29; the fourth, 31 $\frac{1}{4}$; the fifth, 33 $\frac{1}{2}$; the ſixth, 36; and ſeventh, 38.

Theſe experiments being made in preſence of many officers of Artillery, convinced all preſent that none of *Maigrigny*'s principles were true, and that no dependance could be had on his experiments. Yet ſtill the miners are in doubt, whether the diameter of the excavation can be made greater than double the line of leaſt reſiſtance, or if the pit or excavation will not become like a well when overcharged.

The next thing taken into conſideration, was to find the true figure of the excavation, which we meaſured accordingly, as exact as poſſible, and found it to be a paraboloid very nearly; and it is according to this figure that Mr. *de Valliere* has conſtructed his tables, which are now uſed by all the miners in *France*.

But

But as he ſuppoſes the diameter of the baſe to be always equal to twice the line of leaſt reſiſtance, they are far from being ſo complete as he imagin'd, ſince they contain but one caſe only; as will be ſhewn more at large hereafter.

But before I proceed, it will be neceſſary to obſerve, that tho' I ſaid that the figure of the excavation was a paraboloid, yet the quantity of earth to be blown up ſhould be eſtimated by the part ALMB, cut off by a plane, Fig. 3. LM paſſing through the focus or center C of the chamber, parallel to the horizon AB; the other part LEM, being occaſioned by the force of the powder preſſing downwards: It is plain, that ſince the exploſion of gun-powder acts every where, and on all ſides alike, it muſt condenſe the ſolid under the chamber from L to M, by its preſſion downwards, ſo long as it preſſes the earth above LM upwards; and it cannot be ſaid, that any particle of earth under the horizontal line LM can be drove upwards; conſequently, the part ALMB of the paraboloid, will be taken hereafter to expreſs the quantity of earth to be blown up.

To find the content of this ſolid I ſhall premiſe a few of the principal properties of the parabola: 1. That the ſquare of any applicate, as DA, is always equal to the rectangle made by the parameter and the correſponding abſciſſa ED; 2. That the ordinate LM, paſſing through the focus C, is equal to the parameter; 3. And that if in the axis ED, produced downwards to K, ſo that EK be equal to EC, or equal to the fourth part of the parameter, the line CA drawn from the focus to the extremity of any ordinate, is always equal to the correſpond-

ing

ing part KD of the Axis, between that ordinate and the point K.

Hence, becauſe of the right angled triangle CDA, the ſum of the ſquares of CD and DA, is equal to the ſquare of CA or its equal KD; and therefore the ſquare root of the ſum of the ſquares of CD and DA, is equal to the line KD. Whence if CD and DA, are given, the line KD and conſequently CK or its equal CL will be given likewiſe: which ſhews that if the line of leaſt reſiſtance CD, and the radius DA of the baſe are given, the parameter may be found.

The ſolid content of the paraboloid is equal to half the cylinder of the ſame baſe and altitude: if r expreſſes half the circumference whoſe radius is unity; that is if $r = 1.57$.

Then becauſe unity is to 2r, or the radius to the circumference as the ſquares of the radii CL, DA, are to the areas of their circles; we have $r\text{ED} \times \overline{\text{AD}}|^2$, for the ſolid content of AEB, and $r\text{EC} \times \overline{\text{CL}}|^2$, for the ſolid LEM; therefore their difference $r\text{ED} \times \overline{\text{AD}}|^2 - r\text{EC} \times \overline{\text{CL}}|^2$, will expreſs the ſolid required.

But if p expreſſes the parameter LM; then will $p \times \text{ED} = \overline{\text{AD}}|^2$, and $p \times \text{CE} = \overline{\text{CL}}|^2$; theſe values being ſubſtituted into the expreſſion of the ſolid gives $rp \times \overline{\text{ED}}|^2 - rp \times \overline{\text{EC}}|^2$; but ſince $\text{ED} = \text{EC} + \text{CD}$; this value of ED being ſubſtituted, gives $rp \times \text{CD} \times \overline{\text{CD} + 2\text{EC}}$, or becauſe $\text{KD} = \text{CD} + 2\text{EC}$, the expreſſion of the ſolid is reduced to $rp \times \text{CD} \times \text{KD}$.

It is to be obſerved, that becauſe r is a conſtant number, it may be neglected in comparing the ſolids; and ſo $p \times \text{CD} \times \text{KD}$, will be the expreſſion of of the ſolid. And when two excavations are compared together, which have the ſame line of leaſt reſiſtance CD; the ſolid will be expreſſed by the rectangle $p \times$ KD, which ſhould be obſerved becauſe it renders the computation ſhorter and eaſier. Hence, if this ſolid, or the quantity of earth to be raiſed, and the line CD, of leaſt reſiſtance, are given, the parameter p may be found; and having the parameter, and the line CD, the equation $p \times \text{ED} = \overline{\text{AD}}|^2$, will give the radius AD of the baſe.

For if CD $= c$, and the given Solid ALMB $= a$, then becauſe CD $= \frac{1}{4} p$, the Expreſſion $p \times \text{CD} \times \text{KD}$, will give $pc \times \overline{\frac{1}{2}p + c} = a$, or $ppc + 2pcc = 2a$, and dividing by c, $pp + 2pc = \frac{2a}{c}$, to which adding cc, we ſhall have $pp + 2cp + cc = \frac{2a}{c} + cc$; and the ſquare root will be $p + c = \sqrt{\frac{2a}{c} + cc}$.

In the comparing of Mines together, which have the ſame line of leaſt reſiſtance, the rectangle $p \times$ KD gives $pp + 2cp = 2a$; to which adding cc, there will be $pp + 2cp + cc = 2a + cc$ whoſe ſquare root is $p + c = \sqrt{2a + cc}$.

By means of theſe equations, all the different problems relating to Mines are eaſily ſolved, upon this condition, that we ſuppoſe the forces of the powder are proportional to their quantities, and therefore the charges are alſo proportional to the quantities of earth to be raiſed in the ſame ſort of ſoil; that is, in ſoil of the ſame denſity and tenacity.

Many authors aſſert, that the elaſtic force of powder is greater, in proportion, in larger quantities

tities than in ſmall ones; that is, that the force of two pounds is more than double the force of one; but as the only argument they offer to ſupport this aſſertion, is that the heat increaſes in a greater degree, in large quantities than in ſmall ones; and produce no experiments to confirm it: whereas I have proved in my treatiſe of artillery, page 31, corollary 2. That when the ſpaces, in which the powder is confined, are as the quantities, the elaſtic force of powder are equal; that is, they will impel bodies, which weigh in proportion to the quantity of powder, with the ſame velocity: conſequently a double quantity of powder will raiſe a double quantity of earth, and ſo on in proportion to the quantities.

Mr. *Belidor* gives another reaſon for diminiſhing the charges of mines as the earth to be raiſed increaſes, which is, that not only the weight of the earth to be raiſed, is to be conſidered, but likewiſe the preſſure of the atmoſphere over the ſurface of the excavation, which preſſure being as the baſe of the excavation, and this being as the ſquares of the diameters; whereas the weights of ſimilar ſolids are as the cubes of theſe diameters; and therefore this preſſure being leſs in proportion, in larger bodies than in ſmaller, the charges ought rather to be leſſened in large mines than in the ſmall. But how far this reaſon will hold good has not yet been found by experiments; and ſince thoſe experiments he made at *la Fere*, manifeſtly contradict this Hypotheſis, as will be ſhewn hereafter, we may venture to conclude, that notwithſtanding all that has been ſaid hitherto, in favour of larger mines requiring leſſer charges, in proportion to the ſmall, is intirely groundleſs, and confirm that quan-

tities

tities of powder, will always raiſe proportionable quantities of earth.

To know the quantity of powder neceſſary for blowing up of a Mine in a particular ſoil; ſeveral Mines are to be made in it, having their Lines of leaſt reſiſtance equal, but loaded with different quantities of powder, till one is found to have the deſired effect.

This being found, the diameter of its baſe muſt be meaſured with great accuracy, as likewiſe the Line of leaſt reſiſtance ſhould have been very nicely determined; for the leaſt miſtake in this Mine, which is to ſerve as a model to determine thereby the different parts of all the Mines made in the ſame ſoil, would cauſe conſiderable errors; and when theſe Lines are determined, the parameter p, of the parabola, is found by the equation $KD = \sqrt{\overline{AD}^2 + \overline{CD}^2}$: and having the parameter given, the quantity of earth or ſolid is found by the ſolid, $p \times CD \times KD$, or by the rectangle $p \times KD$, according as the lines of leaſt reſiſtance are different or the ſame. This ſolid, and the charge of the Mine, will ſerve to find the effect of any other Mine made in the ſame ſoil when the charge is given; or to determine the charge, ſo that the diameter of the baſe ſhall be of any given length, by means of the equality $p \times ED = \overline{AD}|^2$. The ſame being performed in all the different ſoils, which generally occur in making Mines, will ſerve to make Mines of any depth, or placed in any ſoil.

The miners divide the different ſoils into five ſpecies.

1.	Into	looſe earth or ſand.
2.		common midling light ſoil.
3.		loom or ſtrong ſoil.
4.		potters clay, or ſtiff ſoil.
5.		clay mixt with ſtones.
6.		all kind of maſonry.

It has been found, that a cubic foot of the firſt weighs 95 lb; of the ſecond, 124; of the third, 126; of the fourth, 135; and 160 of the fifth. But as to maſonry, it cannot be determined to any degree of exactneſs, as depending on the different kinds of ſtones or bricks of which they are made.

It is pretended, that there is nine pounds of powder required to raiſe a cubic toiſe of the firſt kind; 11 of the ſecond; 13 of the third; 15 of the fourth; 18 of the fifth; and 20 or 25 to raiſe a cubic toiſe of maſonry above-ground, and 35 or 40 for raiſing the ſame quantity under-ground.

Theſe are the *French* weights and meaſures, which being reduced into *Engliſh,* give 8 lb. of powder for the firſt kind of ſoil; 9.8 for the ſecond; 11.6 for the third; 13.4 for the fourth; 16 for the fifth; 18 or 22.3 for the maſonry above-ground, and 31 or 35 for raiſing the ſame quantity under-ground.

In the ſecond volume of M. *Vauban*'s attac and defence of places, he ſays, that the following rules never fail.

A cubic toiſe of common earth requires 14 pounds of powder to be raiſed.

Stiff ſand or loom, which may be dug without being ſupported, requires 17 pounds *per* toiſe.

Mixt earth requires 18 pounds *per* toiſe.

Potters clay or ſtiff ſoil, 19 pounds *per* toiſe.

Fat or ſtiff earth mixt with pebble ſtone, 22 lb.

Whet

Whet ſand, which cannot be dug without being ſupported, 15 lb.

Theſe rules of Mr. *Vauban*, make therefore the charges greater than thoſe of later miners. It will not be an eaſy matter to decide which of them are beſt, unleſs by experiments; I would therefore adviſe the miners always to make a Mine by way of experiment, before they make any for real uſe.

Suppoſing theſe experiments to be made with great care and exactneſs, it will be no difficult affair to find the proper charge of a Mine, ſo as the diameter of its baſe be of any given length; or when that length is given, to determine the charge required.

For example, let it be required to find the diameter of a Mine made in the ſecond ſort of ſoil; which being loaded with 100 lb. of powder, ſay, if 11 pounds raiſe a cubic toiſe, or 216 cubic feet of earth, how much will raiſe 100 lb.; the fourth term, which is 1964, will be $= a$; and $2a = 3928$; and ſuppoſing the line of leaſt reſiſtance CD to be 10 feet, then will $c = 10$; hence the equation $c + p = \sqrt{\frac{2a}{c} + cc}$, will give $c + p = \sqrt{402.8} = 20$; or $p = 10$, and $EC = \frac{1}{4}p = 2.5 : ED = 12.5$; whence the equation $p \times ED = \overline{AD}|^2$ gives $10 \times 12.5 = 125 = \overline{AD}|^2$, or $AD = 11.2$ nearly.

But to ſhew how far this theory agrees with the experiments mentioned before, I ſhall ſuppoſe the firſt to be true, and from thence proceed to find what the diameters of the baſes of the others will be. All the lines of leaſt reſiſtance of theſe Mines were 10 feet each, the diameter of the baſe of the firſt

firſt Mine was found to be 22 $\frac{2}{3}$ feet; ſo that AD is = 11.33, or = 11.4. CD = 10; theſe values being ſubſtituted into the equation $KD = \sqrt{CD^2 + AD^2}$, will give $KD = \sqrt{229.96} = 15.16$; and 2 KC = p = 10.32; hence theſe values being ſubſtituted into the rectangle $p \times KD$, becauſe the line of leaſt reſiſtance is here always the ſame, gives $10.32 \times 15.16 = 156.5$, for the ſolid, which muſt be remembered, becauſe it is the ſtandard number whereby the other ſolids are determined.

Now if 120 lb. gives 156.5, how much gives the charge 160 lb. of the ſecond for its ſolid, the fourth term gives 208 $\frac{2}{3}$ = a, and $2a$ = 417 $\frac{1}{3}$; this value, as well as that of $c = 10$, being ſubſtituted into $p + c = \sqrt{2a + cc}$, gives $p + c = \sqrt{517.4} = 22.7$; hence $p = 12.7$, $EC = \frac{1}{4}p = 3.2$ and ED = 13.2. Now theſe values being ſubſtituted into $p \times ED = \overline{AD}|^2$, give $\overline{AD}|^2 = 167.64$ and AD = 13 nearly; and as AB has been found by meaſurement to be about 26, it ſhews that this computation anſwers very nearly the experiment.

If as the charge of 120 of the firſt is to the charge 200 of the third, ſo is the ſolid 156.5 of the firſt to the ſolid of the third; we ſhall have $a = 260.84$ or $2a = 521.68$; and as $c = 10$, the equation $p + c = \sqrt{2a + cc}$ gives $p + c = \sqrt{621.68} = 24.93$ nearly; hence $p = 14.93$, $\frac{1}{4}p = EC = 3.73$, and ED = 13.73; theſe values being ſubſtituted into $p \times ED = \overline{AD}|^2$, gives AD = 14.32, and AB = 28.64; which anſwers nearly the experiment; it was found that AB = 29 nearly.

If we proceed thus with regard to the 4th, 5th, 6th and 7th Experiments, we ſhall find the diameters of the baſe to be as follows; that of the 4th,

31.2

31.2; that of the 5th, 33.2; the 6th, 35.3; and that of the 7th, 37.4; which anſwers pretty near the experiments; the greateſt difference does not exceed ſix inches, or thereabouts; which may ariſe from various cauſes, as from the line of leaſt reſiſtance being not exactly ten feet; for a few inches more or leſs may make an alteration; the inequality of the ſoil, which may be more or leſs denſe, and many other accidents that may intervene in the practice, ſo as not exactly to anſwer the theory.

It may not be improper to mention ſome of the advantages this method has above that which is uſed amongſt miners; one of the principal is, that if it be required to make a Mine of a large opening, in a low ground, where it cannot be dug very low without meeting with water, then they are obliged to make two or more Mines near one another, ſo as their joint effects ſhall make the opening required; whereas any opening may be made according to the foregoing principles, let the line of leaſt reſiſtance be what it will.

There are many caſes, eſpecially in the Defence of Places, where a great excavation furniſhes a convenient lodgment to the Beſiegers, which ſhould be avoided, if poſſible; therefore the buſineſs is to make Mines which ſhall overthrow the Beſiegers works, without affording them much covering. This is done by making the line of leaſt reſiſtance but ſmall, and loading the Mine with more powder.

There are other advantages in making the lines of leaſt reſiſtance but ſmall; the ſhafts and galleries are much ſooner made; and ſeveral Mines may be placed under one another, by which the ſame ſpot of ground may be blown up ſeveral times;

times; and as the Befiegers lodge in the hole, fo foon as the Mine is fprung, thinking themfelves fecure, the fecond Mine will do them more mifchief than the firft; and the fame may be faid of the third.

As the places where batteries are made, to make breach in the face of any work, are always known, Mines may be made under them, fo as not only to deftroy thefe batteries, but alfo to make the guns fall in the ditch, and that feveral times, as at *La Fere,* where the fame battery, oppofite to a lunet, confifting of two 24 pounders, was blown up three times, one after another, and the guns every time fell into the ditch.

Suppofe it was required to make a Mine, in the fame fort of foil as that in which the feven experiments mentioned before were made, fo that the line of leaft refiftance fhall be equal to the radius of the bafe, and each of 10 feet, and to find the quantity of powder neceffary for its charge.

Becaufe $AD = CD = 10$, the equation $KD = \sqrt{AD^2 + CD^2}$, will give $KD = \sqrt{200} = 14.14$; hence $p = 8.28$; thefe values being fubftituted into $p \times KD$, will give 117 nearly, for the folid; then if we fay, as the folid 156.5, of the firft experiment, is to the folid 117, fo is the charge 120 to the charge required, which is 90 lb. nearly. But if it was required to find the quantity of powder neceffary to raife a cubic fathom, or 216 cubic feet of this foil; then becaufe $CD = 10$ has been neglected in the folid 156.5, of the firft experiment, as likewife the ratio r, therefore, the quantity muft be multiplied by $r \times 10$; or becaufe $r = 1.57$ by 15.7, which will give 2457; then if we we ſay, as 2457 requires 120 lb. of powder, how much will 216 require; and the fourth term, which is 10.5 lb. will be the number ſought. From whence it appears, that the ſoil, in which theſe experiments were made, was a light ſort of ſoil, ſomewhat lighter than that which is taken by the miners for the ſecond ſort.

Mr. *De Valliere* ſuppoſes, in his Table, which I ſhall inſert here, that a Mine, whoſe line of leaſt reſiſtance and radius of the baſe being each 10 feet, requires 93 $\frac{3}{4}$ lb. for its charge. Now, if it be required to find what kind of ſoil theſe Mines are made in, by ſubſtituting the number for CD, AD, into the equation $KD = \sqrt{AD^2 + CD^2}$, we ſhall have $KD = \sqrt{200} = 14.14$, and $p = 8.28$; now theſe values being ſubſtituted into $rp \times CD \times KD$, we ſhall have $15.7 \times 8.28 \times 14.14 = 1838$; then if we ſay, as 1838 is to 93 $\frac{3}{4}$, ſo is 216 to 11. This fourth term will expreſs the number of pounds of powder required to raiſe a cubic fathom of the ſame ſort of ſoil, which therefore is the ſecond ſort.

The preceding computations have been made of *French* weights and meaſures, to ſhew how nearly the foregoing theory agrees with the experiments made at *La Fere.* It remains now to apply it to our own weights and meaſures; becauſe eight pounds of powder will raiſe a cubic fathom of earth of the firſt ſort; if we ſay a cubic fathom or 216 cubic feet is to eight pounds, 1838 cubic feet is to 68.074 pounds, this fourth term will be the charge of a Mine, whoſe line of reſiſtance is 10 feet as well as the radius of the baſe: in the ſame manner are found the charges of the ſame Mine in the reſt of the ſoils. But the ſhorteſt way of computing

computing tables is to ſubtract the logarithm of 216 from that of 1838, which gives 9298917; now if to this logarithm we add thoſe of 8, 9.8, 11.6, 13.4, 16; the weight of the powder required to raiſe a cubic fathom of the different ſoils, found before; we ſhall have 1° 83298, 2° 92112, 3° 99435, 4° 05699, 5° 13401, for the logarithms of the charges of a Mine whoſe line of leaſt reſiſtance is 10 feet and the diameter of the baſe 20.

VALLIERE'S TABLE *for the Charges of Mines.*

Length of the Line of leaſt Reſiſtan.	Charge of Powder	Length of the Line of leaſt Reſiſtan.	Charge of Powder	Length of the Line of leaſt Reſiſtan.	Charge of Powder	Length of the Line of leaſt Reſiſtan.	Charge of Powder
Feet	lb. oz.	Feet	lb. oz.	Feet	lb. oz.	Feet	lb. oz.
1	0 2	11	124 12	21	868 3	31	2794 4
2	0 12	12	162 0	22	998 4	32	3072 0
3	2 8	13	205 15	23	1140 10	33	3369 1
4	6 0	14	257 4	24	1296 0	34	3680 12
5	11 11	15	316 4	25	1558 9	35	4019 8
6	20 4	16	384 0	26	1647 12	36	4374 0
7	32 2	17	460 9	27	1815 4	37	4748 11
8	48 0	18	516 12	28	2058 0	38	5144 4
9	68 5	19	643 0	29	2286 7	39	5561 2
10	93 12	20	750 0	30	2530 4	40	6000 0

By this conſtruction the radii of the baſes being always equal to the lines of leaſt reſiſtances, the ſolids

ſolids are ſimilar, and therefore are to one another as the cubes of their axis; that is, as the cubes of the lines of leaſt reſiſtances. So that taking any one of the charges to be true, the others will be found by ſaying, as the cube of the axis whoſe charge is given is to its charge, ſo is the cube of the axis of any other Mine to its charge.

For example, let the charge 93 $\frac{3}{4}$ of the Mine, whoſe line of leaſt reſiſtance is 10 feet, be given; and it be required to find the charge of any other Mine whoſe line of leaſt reſiſtance is given, ſuppoſe 15: then ſay, as the cube 1000 of 10 is to the cube 3375 of 15, ſo is the charge 93 $\frac{3}{4}$; to the charge required which is 316.4, or 316 lb. 6 ounces, which is 2 ounces more than in the table. In the ſame manner is found the charge of a Mine whoſe line of leaſt reſiſtance is 20; or becauſe 20 is double of 10, the cube of 20 will be octuple the cube of 10; and therefore $8 \times 93\frac{3}{4}$, or 750 lb. will be the charge for that Mine.

A TABLE

A TABLE *of the* CHARGES *of* MINES *according to the* AUTHOR'S *Theory.*

Diameter.	Charge.	Diameter.	Charge.	Diameter.	Charge.
Feet. 22	lb. 150	Feet. 42	lb. 639	Feet. 62	lb. 1518
24	181	44	711	64	1621
26	217	46	773	66	1741
28	255	48	857	68	1842
30	297	50	946	70	1980
32	344	52	1020	72	2098
34	394	54	1115	74	2243
36	452	56	1205	76	2372
38	502	58	1299	78	2501
40	560	60	1406	80	2648

In this table the line of leaſt reſiſtance is ſuppoſed to be always 10 feet, and the charges producing the openings at the ſides of them from 22 feet to 80. I ſuppoſe that the charge 93 $\frac{3}{4}$ of a Mine, whoſe line of leaſt reſiſtance and radius of the baſe are each 10 feet, be given, and from thence all the reſt are computed by means of theſe equations, $KD = \sqrt{AD^2 + CD^2}$, and $p \times KD = a$; and as I have obſerved, by comparing the diameters of the baſes found by means of theſe equations, to be rather leſs than thoſe found by experiments, it is preſumed that the diameters marked in this table will not be found leſs, but rather greater in practice.

If

If we compare the charges of both tables, which ſhould produce the ſame diameters, or openings, it will be found that it requires 190 lb. leſs in my table to make a diameter of 40 feet, 1124 to make one of 60, and 3352 leſs to make one of 80 feet, which plainly ſhews the great advantages of this theory above that commonly uſed; and the great quantity of powder ſaved thereby. If there were ſuch experiments made as could be depended upon in different ſoils, tables might be conſtructed accordingly, and for any depths, from 5 feet to 20, there need be no greater; but as there are no ſuch experiments, we muſt content ourſelves, for the preſent, with that above, which may ſerve as a pattern to any one who has a mind to conſtruct others.

Having found, by the foregoing method, the quantities of powder with which Mines are to be loaded; it remains now to ſhew, how to find the bigneſs of the box to contain it, which are generally made cubical; and as a cubic foot of common powder weighs about 55 pounds, if we ſay as 55 is to unity, ſo is any other quantity to its cube; that is, if the given quantity of powder be divided by 55, the quotient will be the cube required; and its cube root will be the length of the ſide of the box.

For example, to find a cube which ſhall hold 360 lb. divide 360 by 55; you will have 6.545 for the cube expreſſed in feet, and the cube root 1.875 feet, or 22.5 inches of that number will be the length of the ſide of that cube.

Tho' the chambers are uſually made cubical, yet when the quantity of powder is pretty conſiderable, it will be more advantagious to make them flat,

flat, as the bottom of the excavation will afford a larger lodgment, as likewife the line of leaft refiftance, which is taken from the center of the chamber, will be more proportionable to the earth above it.

It would, in my opinion, be proper to make all the chambers of the fame height with that of the Mine made for the experiment. In that cafe, the fide of the chamber will be found by dividing the given quantity of powder, by the product of 55 multiplied by the given height, and extracting the fquare root, which will be the length of the fide required.

For example, to find the fide of the box which fhall hold 360 lb. of powder, whofe height is a foot; then becaufe 55 × 1 is 55, and 360 divided by 55, gives 6.5455; whofe fquare root 2.558 feet, or 30.7 inches will be the length of the fide required.

N. B. It is to be obferved, that the box muft always be made a fourth bigger than it fhould be, on account of the ftraw and fand-bags put in it, to keep the powder from receiving any damage from wet; fo that if the quantity of powder be 360 lb. the fourth part of it, which is 90, muft be added, and the fum 450 divided by 55, to have 8.1818, whofe fquare root 2.86 feet, or 34 inches, will be the fide required.

It may not be improper to obferve likewife, that when a chamber happens to be placed on a rock, or any other hard fubftance, the force or action of the powder downwards, meeting with a great refiftance, will be employed in raifing the earth upwards; and confequently the effect of the Mine will be much greater than that produced by the fame

quantity

quantity of powder, being placed on a ſofter ſubſtance. Whence if a platform of ſtrong planks or ſtones were made under the chambers, there would be a leſſer quantity of powder required for the charge of the Mine.

If veſſels of caſt iron were made in the form of the fruſtrum of a cone, pretty large above, ſo as to hold the neceſſary charges of Mines, they may be of great advantage in places beſieged; for if they are placed on a ſtrong platform, and ſtrongly buttreſſed above againſt the earth, there need be but little of the gallery ſtopped up; and when the Mine is ſprung, the miners may enter the gallery and remove the veſſel a little nearer, and load it again; and this may be done ſeveral times, ſo that many Mines may be ſprung in the ſame gallery with very little trouble.

A TABLE *ſhewing the dimenſions of the cube boxes, for holding the charges of Mines, from* 50 *to* 640 *pounds, expreſſed in inches and tenths.*

Charge.	Side.	Charge.	Side.	Charge.	Side.	Charge.	Side.
50	12.5	150	18.0	250	21.4	450	26.1
55	12.9	155	18.3	260	21.7	460	26.2
60	13.3	160	18.5	270	22.0	470	26.4
65	13.6	165	18.6	280	22.2	480	26.6
70	14.0	170	18.8	290	22.5	490	26.8
75	14.3	175	19.0	300	22.8	500	27.0
80	14.6	180	19.2	310	23.0	510	27.2
85	14.9	185	19.4	320	23.3	520	27.3
90	15.2	190	19.6	330	23.5	530	27.5
95	15.5	195	19.7	340	23.7	540	27.7
100	15.8	200	19.9	350	24.0	550	27.9
105	16.0	205	20.0	360	24.2	560	28.0
110	16.3	210	20.2	370	24.4	570	28.2
115	16.5	215	20.4	380	24.6	580	28.4
120	16.8	220	20.5	390	24.8	590	28.5
125	17.0	225	20.7	400	25.0	600	28.7
130	17.2	230	20.8	410	25.3	610	28.8
135	17.4	235	21.0	420	25.5	620	29.0
140	17.6	240	21.1	430	25.7	630	29.1
145	17.8	245	21.3	440	25.9	640	29.3

The

The *French* authors pretend that a cubic foot of powder weighs 70 pounds, which is certainly not true, at leaſt with regard to the powder made at preſent. For I found by repeated trials, that the cube of the axis of an equilateral cylinder, which holds a pound of powder, was 40 cubic inches; and as the content of the cylinder is to that of the circumſcribed cube as 355 is to 452, and the cube 40 is to the cubic feet or 1728 cubic inches; as the weight found by the laſt proportion is to the weight of a cubic foot of powder, which will be found to be 55 pounds.

The *French* foot being to our foot as 16 to 15 nearly, and their pound to ours as 68 to 63; the cube of 16 is to the cube of 15 as 1728 of our inches is to 1424 of theirs; and 68 is to 63 as 55 pounds *Engliſh* is to 50.9 pounds *French*: therefore 1424 is to 1728 as 50.9 to the weight of a *French* cubic foot, which will be found to be 61.7 pounds nearly.

The manner of directing the GALLERIES.

SUppoſe a ſhaft is to be ſunk at A, and the Mine to be carried under an object B, and to be of a certain depth; the firſt thing to be done is, to find the diſtance AB exactly, either by trigonometry, or by any other means; then the bearing of this line AB with ſouth and north, muſt alſo be found exactly, by means of a good needle and compaſs, which we ſuppoſe to be repreſented by the angle NAB; then the ſhaft is ſunk at A, of about 5 or 6 feet each way, ſo low as the Mine is to be deep; then the miner directs his gallery according to the given bearing, uſing a Fig. 4.

maſon's

mafon's level to carry it on, horizontally; and as the gallery is carried on, he meafures it, till he finds himfelf under the object B, where he places his chamber.

If the object B is either higher or lower than the place A, where the fhaft is made, then the excefs of the object above or under the place A muft be found, and either fubtracted or added to the depth of the fhaft, and then the gallery is carried on as before.

It may happen that there is an obftacle in the way, fuch as a rock, water, or a morafs; in fuch cafes the gallery muft be made with turnings at right angles to each other, fuch as from C to D till it is clear off from the obftacle; after this, it is made from D to E parallel to the firft part AC; and at E, it is made perpendicular to DE and EF equal to CD; by which means the point F will be in the firft direction, in which it is carried on till the fum of the three parts AC, DE, FB, is equal to the given diftance of the object B, from the fhaft A.

Fig. 5.

Various cafes may happen in the carrying on the galleries, the particulars can neither be forefeen nor defcribed, for which reafon I fhall leave them to the skill and knowledge of experienced miners.

It is unqueftionably of great importance to make the Mine exactly under the object to be blown up; for want of which, much time, labour, and expence will be loft to no purpofe; it is likewife neceffary to know pretty near how deep the Mine is under the object, that its charge may be rightly determined; if that cannot be known fufficiently near, it will be better to overcharge the Mine than put too little powder in it.

The

The reader may eafily perceive that a good miner ſhould be perfectly well skill'd in the practical part of geometry, without which he will never be able to conduct Mines with any degree of certainty; and therefore I would adviſe thoſe who are deſirous to underſtand this buſineſs well, to ſtudy geometry firſt.

MINES *of different ſorts.*

IF a Mine has but one chamber, ſuch as A, Fig. 6. it is called a ſingle Mine; if it has two chambers, as Fig. 7, it is called double; and if it has three, triple, and ſo on; their names are taken from the number of chambers.

If a ſingle Mine is made under the rampart, to make Breach, the entrance O muſt not be oppoſite to the place where the chamber is deſigned to be, on the contrary, it ſhould be made on one ſide or other; and the gallery with two turnings, as in *Fig.* 6. that it may be ſtopped with more ſecurity; and that the diſtance of the entrance O to the chamber A may be greater than the length of the line of leaſt reſiſtance, otherwiſe the Mine would have its effect that way: it muſt alſo be obſerved, that the chamber is placed in the middle of a counter-fort, by which means it will make a greater Breach than if it were placed in the earth behind the wall.

When a double Mine is made under the rampart for making Breach, the entrance O, *Fig.* 7. is made as near as can be gueſs'd, in the middle, between two counter-forts; the gallery being carried quite through the wall in a direct line, turns afterwards to the right and left, in the form of a T; from whence it is alſo called a T Mine; and the chambers are alſo placed in the next counter-forts; but exactly at equal diſtances from the direct gallery: this

this double Mine will make a much larger Breach than the ſingle one, and it is for that reaſon preferred to any other.

But when a triple Mine is to be made under the rampart, the opening O, *Fig.* 8. is to be made directly oppoſite the counter-fort, if poſſible, and carried directly through the wall, and turned to the right and left in the ſame manner as the former; and the chambers A, B, at both ends, are placed in the two adjacent counter-forts. As to the gallery of the third, C, it is carried round the middle counter-fort, and the chamber placed under its extremity; this laſt is generally charged with fifty pounds of powder more than either of the others: but great care muſt be taken to carry the auget of this laſt chamber in Zig-zagues, ſo as to be equal in length to that of the chamber B, otherwiſe the fire would not reach them all three at the ſame time, and thereby the chamber C not take fire, which ſometimes happens, and then the effect does not anſwer the expectation.

There are ſeldom or ever more made than a triple under the ramparts in Sieges; but when a work is to be demoliſhed, they make then as many as will demoliſh a whole face at once; which is done by giving the fire to all at the ſame time; that is, all the ſauciſſons are brought into one, as ſo contrived, as their parts from the chambers to the common junction be exactly equal.

REMARK.

THough we have ſuppoſed in the Figures 7 and 8, that the entrance O into the galleries is exactly in the middle of the extreme Mines, yet the reader may eaſily ſee that it is impoſſible in prac-

tice

tice, ſince, neither the places nor the diſtances between the counter-forts, can be ſeen or known from the outſide of the wall: therefore it can only be gueſs-work; but when the wall is pierced, the gallery muſt be carried to the right and left, till the counter-forts are found; and then, if the diſtance to one is greater than the other, the auget and ſauciſſon of the neareſt muſt be made in a Zig-zague, ſo as to make it equal in length to that which is fartheſt. In all places fortified by Mr. *Vauban*, the diſtance from the center of one counter-fort to that of the next, is always 18 feet, when the revetement is as high as the rampart, or 15 feet when it is not ſo high; ſo that when one counter-fort has been found, the diſtance to the next is known; but in any other fortification this obſervation is uſeleſs.

I have thus given a sketch of all the material parts of the art of mining, and ſettled the principles on a juſt foundation, which have not hitherto been done by other authors. They in general tranſcribe the works of Mr. *Vauban* and other practical miners, not conſidering, that tho' their works were of great uſe in their time, when this art was in its infancy, yet they can be but of very little uſe ſince there has been ſo many improvements made; what thoſe wanted is now brought to its greateſt perfection. M. *de Valliere* has moſt judiciouſly treated the method of making countermines under the glacis, beſides what relates to thoſe made by the beſiegers, either to avoid or forward their approaches, with more expedition and ſafety. The experiments made at *La Fere* demonſtrate, that what had been done was trivial to what might be done.

F I N I S.

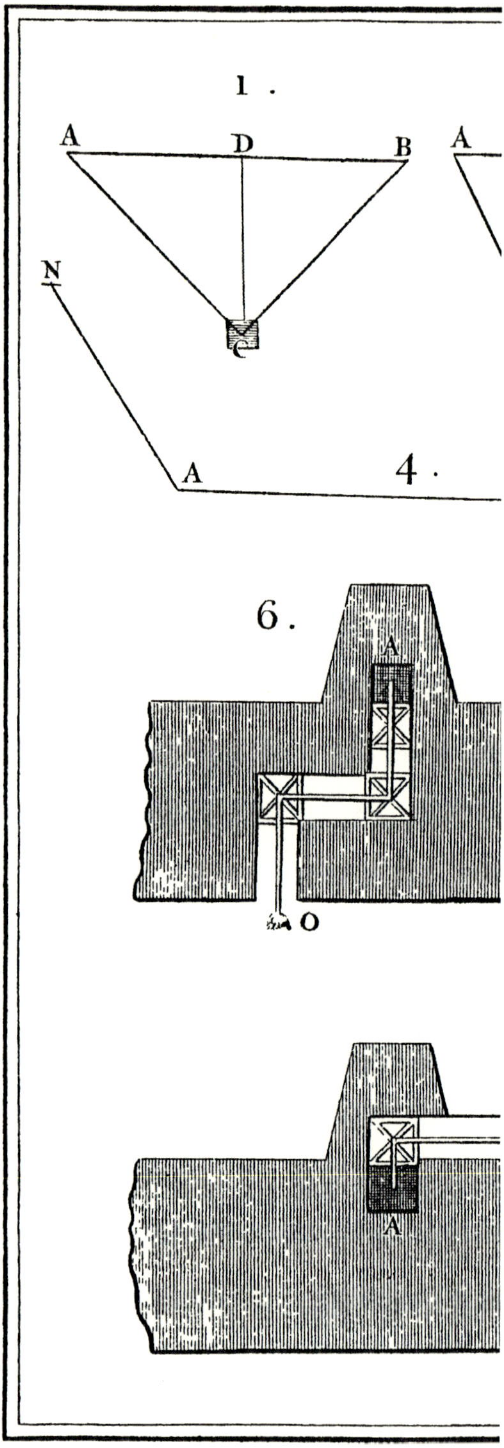
1.
A
D
B
A
C
N
A
4.
6.
A
O
A

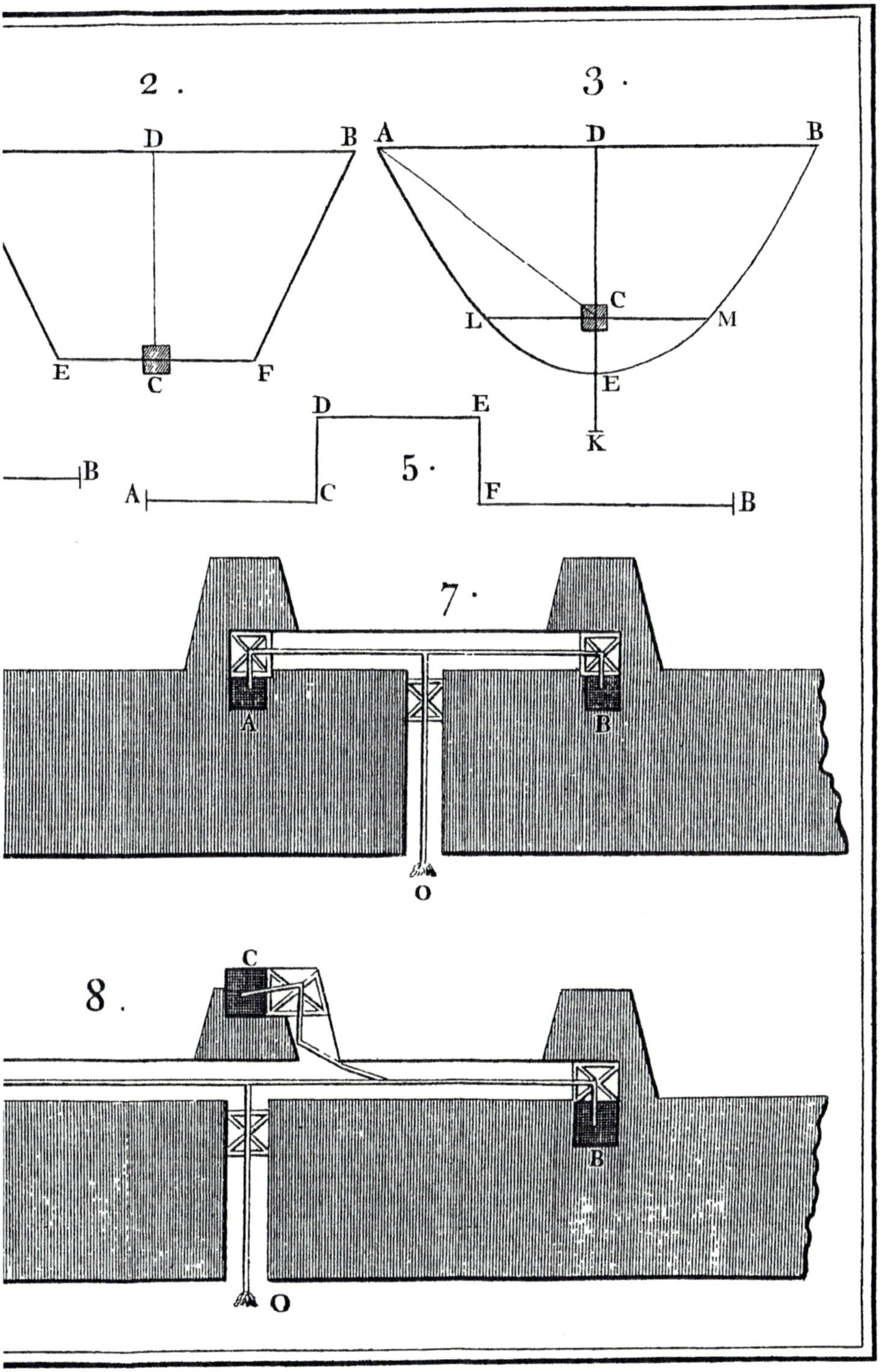
2.
D
B
E
C
F
3.
A
D
B
C
L
M
E
K
B
A
C
D
E
5.
F
B
7.
A
B
O
8.
C
B
O

Just Published for the Use of the Royal Academy *at* Woolwich, *and for all concerned in the Art of War.*

MULLER's Syftem of Mathematics, Gunnery, Fortification, Artillery, &c. in 6 Vol. with above 100 large Copper-Plates, finely engraved, 1 l. 16 s.

Vol. I. containing Algebra, Geometry and Conic Sections.

Vol. II. Trigonometry, Surveying, Levelling, Menfuration, Laws of Motion, Mechanics, Projectiles, Gunnery, Hydroftatics, Hydraulics, Pneumatics, Theory of Pumps, &c.

Vol. III. Elements of Fortification, with Remarks on *Vauban, Coehorn, Bellidore,* &c. fecond Edition improved.

Vol. IV. Practical Fortification, Theory and Dimenfions of Walls, Arches and Timbers; Properties, Qualities and manner of ufing Materials; of facing a Fortrefs, to eftimate the Works, to build Aquatics, as Stone-bridges, Harbours, Quays, Wharfs, Sluices and Aqueducts.

Vol. V. Artillery, general Conftruction of Brafs and Iron Guns for Sea and Land, their Carriages; Mortars and Howitzes, with their Beds and Carriages: Dimenfions of all other Kinds of Carriages in Artillery. Exercife at Home and on Service in a Siege or Battle, its March, Encampment, Ammunition, Stores and Horfes, Laboratory Works, Theory of Powder, &c.

Vol. VI. Attac and Defence of fortified Places: Preparations for and Operations of an Attac; and the Defence of every Part of a Fortification. Making and loading of Mines, with Tables of their proper Charges. The Dimenfions of the Cubic Boxes for holding from 50 to 640 lb. of Powder for their Charges, on a new Theory. The fecond Edition corrected, and very much enlarged with new Tables, &c. by the Author.

A Syftem of Camp Difcipline, Military Honours, Garrifon Duty, and other Regulations for the Land Forces, collected by a Gentleman of the Army; in which are included *Kane*'s Difcipline for a Battalion in Action. With a Map of the Seat of War, Lines and Plans of Battles, and above fixty Military Schemes, finely engraved from the Originals of the moft eminent Generals, &c. To which is added, General *Kane*'s Campaigns of King *William* and the Duke of *Marlborough,* improved from the late Earl of *Craufurd*'s and Colonel *Dunbar*'s Copies, taken from General *Kane*'s own Writing. With his Remarks on the feveral Stratagems by which every Battle was won or loft, from 1689 to 1712. The fecond Edition continued from the Reftoration where our Standing Army commences, in a Series of Hiftorical and Chronological Facts of the Military and Naval Tranfactions of *Great Britain*; being a concife Hiftory (to fupply the Scenes of Action in which the General was not engaged) to 1757. By an Impartial Hand. *London,* printed for *J. Millan,* oppofite the Admiralty Office, 1757. Price 7 s. 6 d.

A Regimental Book for Commiffioned, Non-commiffioned, Staff and Private Men, &c. with all Sorts of Books, Returns, Forms, &c. relating to the Army.

NOTES

ON THE

ATTAC and DEFENCE

OF

FORTIFIED PLACES.

by DAVID MANTHEY.

CONTENTS.

INTRO-

INTRODUCTION *and* BIOGRAPHICAL NOTE

IN the Preface, *John Muller* gives ample reason for the original publication of *The Attac and Defence of Fortified Places.* The reasons presented are equally valid for the publication of this reprint. While siegeworks and fortifications are no longer studied with the same attention and zeal as when they were in principal military use, knowledge of them is required to understand the actions of military commanders, the nature of battles, and the progress of history.

Many readers and historians are familiar with Marshal *Sebastien LePrestre de Vauban*, who formulated the scientific siege, developed numerous tools and weapons, and profoundly affected the military tactics and strategies used in the 17th century. Just as *Vauban* did before him, *John Muller* wrote the principal guides on fortification and siege in the mid 18th century. His writings included books on the engineering and construction of military works, the use of artillery, and, of course, this book on the attack and defense of places. *Muller*'s books had wide circulation and frequent reprints; these texts were not just academic, but were used in the field by the armies of the time.

John Muller was born in Germany in 1699. It is likely that he moved to England at an early age. All of his professional work was done in the service of England, and all of his books were published in London.

His first book was *A Mathematical Treatise, containing a System of Conic Sections and the Doctrine of Fluxions and Fluents applied to various Subjects*, published in 1736 and dedicated to the Duke of Argyll and Greenwich, Master-General of the Ordnance. This book shows an adept knowledge of mathematics, including calculus, and, although some of it appears peculiarly cumbersome

by

by today's standards, it is a well-written and competent work. It was translated into French in 1760.

In July 1741 *Muller* was appointed head-master of the Royal Military Academy in Woolwich with a salary of £100 a year. The academy was improved by the formation of a cadet company in 1746, and, at same time, *Muller* was made Professor of Fortification and Artillery. In November 1754, he received a raise to £200 a year and a promotion to Chief Master of the Academy. He retired from the Academy in September 1766, and was some time later granted a pension of £50 a year.

In 1746, *Muller* published the *Elements of Fortification, with Remarks on Vauban, Coe-horn, Bellidore, &c.* This book was the most influential of all of *Muller*'s works. It contains instructions on how to design fortifications to suit a particular terrain and defensive objective. The book went through at least six editions, the last of which was in 1807.

In 1747, the first edition of *The Attac and Defence of Fortified Places* was published. The second edition was published in 1757; this is a reprint of that edition. A third edition was printed in 1770, and a fourth edition was published in 1791. The work borrows heavily from *Vauban*, which each subsequent edition containing more new material, mainly on mining.

Muller continued to write for the military in general and for the Royal Academy of Artillery in particular. *Elements of Mathematics* came out in 1748. The *Practical Part of Fortification* was published in 1755, and covers what would now be called civil engineering, detailing how to actually build fortifications. *A Treatise of Artillery* was printed in 1757; this work is considered the premiere reference on ordnance from the 18th century. His first six books were published as a set in 1757, clearly with the intent that they would be used as schoolbooks by the cadets in the Academy.

In

In 1760 *Muller* translated and published *The Field Engineer* written by the Frenchman *Louis André de la Mamie de Clairac.* This book neatly complements *Muller*'s own works by providing instructions specifically for the engineering corps. At the Royal Academy and in his own writings, *Muller* principally instructed officers and commanders who had engineers working under them and who needed to be able to understand and direct the engineers' activities.

In 1769, *Muller*'s last book, *A New System of Mathematics*, was published.

All of *Muller*'s works went through multiple editions. Many of the books, including *The Attac and Defence of Fortified Places*, continued to be printed and used by the military establishment after *Muller*'s death.

John Muller died in April 1784. A brief obituary of the time refers to him as the "author of many ingenuous works." A better testament is given in a later commentary: "Mr. *Muller*, of Woolwich Academy, the scholastic father of all the great engineers which this country has employed for forty years."

ABOUT *this* REPRINT

THIS reprint has been typeset to be a faithful copy of the original text. The typeface has been designed to match the original. The spelling and punctuation have been maintained. Even the pagination and placement of words on each line is the same.

A few printer's errors have been corrected. This has only been done where it is obviously an error in the original typesetting, rather than a variation in spelling or layout. Specifically, two errors have been corrected. On page 5 in the description of the *Chevaux de Frize*, the original text read 'staves of aboot an inch and a quarter.'

On

On page 79, the original had a double comma in the first full paragraph, thus: 'f, f, Batteries in the places of Arms,, to fire at the faces.'

At the bottom of most pages is a catchword. This word is intended to be the same as the first word on the subsequent page, and was used to aid both the reader and the bookbinder. The catchwords in this book do not always match the following page; rather, the catchwords have been copied from the original, and any deviations therein have been duplicated.

It should be noted that the original text on page 133 has been maintained, even though there is a discrepancy in the number of mortars listed. The first full paragraph should almost certainly read 'so that the whole will amount to 72 pieces of cannon and 12 mortars.'

Likewise, the table on page 238 does not agree with the premise that a mine with a diameter of 20 feet and a line of least resistance of 10 feet requires a charge of 93 $\frac{3}{4}$ pounds of powder. Specifically, the amount of powder was converted from French Pounds to English Pounds, but the linear measurements were improperly converted between the French and English values. All of the values in the table are roughly forty-five percent larger than expected. See the notes on units on page 254 and the section on the equation of mines on page 258 for more information.

Another mistake is the scale shown in Plate VII. The scale is in feet, though it has been labelled fathoms. There are a few similar problems with the original figures that are given in the text, such as a mathematics error on page 231. All of these have been faithfully copied from the original.

One major error that has been corrected is the page numbering listed in the table of contents. The original book listed page numbers that did not correspond to the text. Universally, the number listed was too

great,

great, with the early pages being only one or two different from the intended location, and the last entries in the table being as much as 15 pages off. It seems probable that the page numbers were erroneously copied based on a galley proof that was then changed extensively to reduce the thickness of the book without changing the content.

In the original book, all of the plates folded out. Each plate is marked with the page that it originally faced. In this reprint, the plates have been split so that they can be printed on the same paper as the text, and they have also been shifted one page in location, in general to the rear of the book. The original book had a page size of roughly 5 3/8 inches wide by 9 inches tall. In order to better accommodate the plates, the page width has been mildly enlarged, and the margins on either side of the text increased. Most of the plates are printed at their original size. Plates VI, XIX, XXIII, and XXIV have been reduced slightly in order to fit within the book.

UNITS *of* DISTANCE *and* MASS

THERE are several units of distance used in *The Attac and Defence of Fortified Places* that may not be familiar to the modern reader. Even an 18th century English reader might find the constant use of the fathom somewhat odd. Some of the choices of units were made because *Muller* based significant sections of the book from the works of the Frenchman *Vauban*. In the French system of units, the toise is extensively used. The toise is equal to 6 French feet, to which the rough equivalent is the English fathom.

The English Foot is identical to the foot used in England and America today. This is also known as the

Statute

Statute Foot, based on long-standing laws of England, or alternately as the Survey Foot, as it was used as the basis for early land surveying in America. Today, another common unit of measure is the International Foot, which is defined as exactly 0.999998 Statute Feet. The International Foot is also defined as exactly 0.3048 meters, and was specifically created in an attempt to harmonize the statute system of measurement with the international metric system (officially known as Le Système International d'Unités, or SI).

The following table shows the conversion of different units of distance. The conversion factor used in this table is taken as 114 English feet are equal to 107 French feet. The conversion used in the 18th century varied to some extent; this conversion factor was one of the most common.

A TABLE

A TABLE *for* CONVERTING DISTANCE

Unit	English	French	Meters
	Equivalents		
1 English Inch		0.93860 Inches	0.02540 m
1 English Foot	12 Inches	0.93860 Feet	0.30480 m
1 English Yard	36 Inches 3 Feet	2.81579 Feet	0.91440 m
1 Eng. Fathom	6 Feet 2 Yards	5.63158 Feet 0.93860 Toises	1.82280 m
1 English Mile	5280 Feet 1760 Yards 880 Fathoms	4955.79 Feet 825.96 Toises	1609.35 m
1 Eng. League	15840 Feet 5280 Yards 2640 Fathoms 3 Miles	14867.4 Feet 2477.89 Toises 1.08584 Leagues	4828.04 m
1 French Inch	1.06542 Inches		0.02706 m
1 French Foot	1.06542 Feet	12 Inches	0.32474 m
1 French Toise	6.39252 Feet 1.06542 Fathoms	6 Feet	1.94845 m
1 Fr. League	14587.7 Feet 2.76283 Miles 0.92094 Leagues	13692 Feet 2282 Toises	4446.35 m
1 meter (m)	39.37 Inches 3.28083 Feet 1.09361 Yards 0.54681 Fathoms	36.9525 Inches 3.07938 Feet 0.51323 Toises	
1 kilometer (km)	0.62137 Miles 0.20712 Leagues	0.22490 Leagues	1000 m

In addition to differences between the English and French distance measurements, there are also differences in the mass standards of the two nations. The English Pound is the same pound used in America and Britain today.

today. The conversion factor is taken as 108 English pounds are equal to 100 French pounds. The conversion factor between English pounds and kilograms is taken from official publications issued by the United States National Institute of Standards and Technology (NIST).

A TABLE *for* CONVERTING MASS

Unit	English	French	Kilograms
	Equivalents		
1 Grain	0.00229 Ounces	0.00247 Ounces	0.00006 kg
1 English Ounce	437.5 Grains	0.92593 Ounces	0.02835 kg
1 English Pound	7000 Grains 16 Ounces	0.92593 Pounds	0.45359 kg
1 French Ounce	472.5 Grains 1.08 Ounces		0.03062 kg
1 French Pound	7560 Grains 1.08 Pounds	16 Ounces	0.48988 kg
1 kilogram (kg)	15432.3 Grains 2.20462 Pounds	2.04132 Pounds	

MATHEMATICAL SYMBOLS

IN the 18th century, basic mathematical notation was still under development and had not been fully standardized. Additionally, geometry was the primary system of mathematics used by engineers and other technical workers. Algebra, while known, was less commonly employed. Thus, algebraic symbols are not completely consistent even within a single work such as this book.

In the algebraic equations, parenthesis ( ) are not used for grouping. Rather, parenthesis are only used as punct-

punctuation marks. To indicate algebraic grouping, a *vinculum*, or horizontal line, is placed above the group. For example, the expression, $rp \times CD \times \overline{CD + 2EC}$, from page 226, would be expressed in modern notation as $rp \times CD \times (CD + 2EC)$.

To clarify exactly what quantity is being raised to a power, a vinculum with a vertical bar is often used to delineate the operand. For example, $\overline{AD}|^2$ indicates the quantity AD squared.

A good table of logarithms was considered essential for any serious computation, since the use of logarithms reduces the process of multiplication to the easier job of addition. For this purpose, the logarithms used are base 10; that is, $\log(10) = 1$. Using base 10 allows a comparatively short table to be easily used for any number, as a shift in the decimal point of a number changes the logarithm by an integer quantity. Often, only the fractional part of a logarithm is listed, with the location of the final decimal place assumed to be obvious.

On page 236, the logarithms of several values are given. The number followed by the ° symbol is an ordinal. That is, 1° means 1st, 2° means 2nd, 3° means third, and so forth. The value given after these ordinals is the factional part of a logarithm. The phrase 1° 83298 indicates that for the 1st condition, the fractional part of the logarithm is 0.83298. This corresponds to the logarithm of 68.0738. Since $68.0738 = 10^{1.83298}$, the logarithm of 68.0738 is equal to 1.83298. The integer quantity 1 has been dropped in the value shown by *Muller*.

The EQUATION *of* MINES

A considerable amount of work is shown to calculate the amount of powder required to blow up a mine with a specified line of least resistance and a specified

radius

radius at the surface above the mine. This explanation may be confusing to the modern reader due to the use of the 18th-century style of geometric mathematics. Additionally, the table of charges by *Valliere* on page 236 is given in French feet and French pounds, whilst the table by *Muller* on page 238 is given in English feet and English pounds, but was incorrectly converted from the French.

Two methods are used to compute the amount of soil that is to be blown up by a mine. *Valliere*'s method uses the volume of a paraboloid. This is the solid of revolution given by the curve ALEMB in Plate XXV, Figure 3. Recall that the volume of a paraboloid is half that of a cylinder with the same height and radius, or, expressed as an equation, $\frac{1}{2}\pi r^2 h$. Let r be the radius at the top of the paraboloid (the distance from A to D in the figure), and let l be the line of least resistance (the distance from C to D), then the height of the paraboloid is given by the equation $h = \frac{1}{2}(\sqrt{r^2 + l^2} + l)$. The volume of the paraboloid is then $\frac{1}{2}\pi r^2 \frac{1}{2}(\sqrt{r^2 + l^2} + l)$.

Muller's method instead uses the volume of the paraboloid above the plane of the charge of the mine. This is the solid of revolution given by the outline ALCMB, and can properly be called the frustum of a paraboloid. The volume of this frustum of the paraboloid is calculated by subtracting the bottom of the paraboloid from the whole paraboloid. The equation is $\frac{1}{2}\pi r^2 \frac{1}{2}(\sqrt{r^2 + l^2} + l) - \frac{1}{2}\pi \frac{1}{2}(\sqrt{r^2 + l^2} - l)^3$, which, when simplified, is equal to $\pi l (r^2 + l^2 - l\sqrt{r^2 + l^2})$.

In both cases the volume of the soil that is to be blown up is multiplied by a factor to account for the amount of powder needed per cubic foot of soil and by another factor to account for the tenacity of the soil. Because of this, the multiplicand of $\frac{1}{2}\pi$ is ignored, and is instead incorporated into these factors. Similarly, when *Muller* computes the amount of powder required on

page

page 238, he has dropped one of the factors of *l* since his computations are all for the same line of least resistance.

The table of charges for mines by *Valliere* uses French feet and French pounds. The table is based on a mine with a line of least resistance of 10 French feet and a radius of 10 French feet requiring 93 $\frac{3}{4}$ French pounds of powder. This is the standard from which all other charges are computed.

To convert this table to English units, it is necessary to find the charge in English pounds required for a mine with both a line of least resistance and a radius of 10 English feet. Since 1 French foot is equal to 1.06542 English feet, the standard mine used by *Valliere* has both a line of least resistance and a radius of 10.6542 English feet. 1 French pound is equal to 1.08 English pounds, therefore this mine requires 101 $\frac{1}{4}$ English pounds of powder. Based on the volume of a paraboloid, a mine with a line of least resistance and radius of 10.6542 feet has a volume of 2293.13 cubic feet, while a mine of 10 feet has a volume of 1896.12 cubic feet. The amount of powder required for a mine of 10 feet is calculated by multiplying 101 $\frac{1}{4}$ pounds by the ratio of the volume of the mines, which gives a value of 83.72 English pounds, which is nearly 83 $\frac{3}{4}$.

Here follows *Valliere*'s table revised for English units of measurement, with 83 $\frac{3}{4}$ English pounds used as the charge of a mine whose line of least resistance and radius of the base are each 10 English feet. The revised table uses the volume of the paraboloid of earth blown up, and the mines all have a radius which is equal to the line of least resistance.

VALLIERE'*s*

VALLIERE'*s* TABLE *for the Charges of Mines, revised for* ENGLISH FEET *and* POUNDS.

Length of the Line of least Resistan.	Charge of Powder	Length of the Line of least Resistan.	Charge of Powder	Length of the Line of least Resistan.	Charge of Powder	Length of the Line of least Resistan.	Charge of Powder
Feet	lb. oz.	Feet	lb. oz.	Feet	lb. oz.	Feet	lb. oz.
1	0 2	11	111 8	21	775 10	31	2495 0
2	0 11	12	144 12	22	891 12	32	2744 5
3	2 4	13	184 0	23	1019 0	33	3009 12
4	5 6	14	229 13	24	1156 12	34	3291 12
5	10 8	15	282 11	25	1308 10	35	3590 13
6	18 1	16	343 1	26	1472 0	36	3907 7
7	28 12	17	411 7	27	1648 7	37	4242 3
8	42 14	18	488 7	28	1838 8	38	4595 8
9	61 1	19	574 7	29	2042 9	39	4968 0
10	83 12	20	670 0	30	2261 4	40	5360 0

If *Muller*'s method of calculating volume using the frustum of the paraboloid is used instead of using the whole paraboloid, the ratio of the volumes is the same, and thus the powder charge for a mine of 10 English feet is still 83 $\frac{3}{4}$ English pounds. Specifically, the volume of the frustum of the paraboloid of a mine of 10.6542 feet is 2225.63 cubic feet, while the volume for a mine of 10 feet is 1840.30 cubic feet.

In *Muller*'s table on charges of mines, he has mistakenly reversed the ratio of the length of the French foot and the English foot. Instead of using a value of 83 $\frac{3}{4}$ English pounds for the powder necessary to blow up a mine of 10 English feet in both radius and

and line of least resistance, he has used a value of 122.45 pounds. When discussing the required powder to blow up a cubic fathom of earth on page 230, and when mentioning the density of powder on page 243, *Muller* did the conversion correctly between English and French units.

Here follows a corrected version of *Muller*'s table of charges of mines, with 83 ¾ English pounds used as the charge of a mine whose line of least resistance and radius of the base are each 10 English feet. The table uses the volume of the frustum of the paraboloid, and the mines all have a line of least resistance of 10 English feet.

MULLER'S TABLE *for the Charges of Mines, corrected for* ENGLISH FEET *and* POUNDS.

Diameter.	Charge.	Diameter.	Charge.	Diameter.	Charge.
Feet.	lb. oz.	Feet.	lb. oz.	Feet.	lb. oz.
22	103 7	42	440 15	62	1051 4
24	125 8	44	489 7	64	1127 11
26	150 2	46	540 12	66	1206 15
28	177 4	48	594 12	68	1289 0
30	206 14	50	651 9	70	1373 15
32	239 4	52	711 3	72	1461 11
34	274 3	54	773 9	74	1552 4
36	311 13	56	838 12	76	1645 11
38	352 2	58	906 12	78	1741 15
40	395 3	60	977 10	80	1841 0

On

On GUNPOWDER

GUNpowder is of central importance to sieges and to most 18th century military action. Due to a number of factors, the strength and quality of gunpowder varied from batch to batch. This affected the charges used in mines and artillery and the quantity of gunpowder required to withstand or conduct a siege. To explain how gunpowder can vary, here follows a description of how it was manufactured and how 18th century gunpowder compares with modern black powder.

It should be obvious that one should not attempt to make gunpowder without observing all laws and safety precautions. What is given here is not sufficient for safe or complete manufacture; numerous people have died while trying to make their own black powder. See the book by *von Malitz* listed in the bibliography for a modern approach and safety instructions.

In the 18th century, what is now called black powder was referred to simply as gunpowder. It is a compound that consists of three ingredients intimately ground together and formed into grains. Gunpowder is fairly stable, and will not explode under relatively harsh treatment, yet ignites easily and with significant power. Depending on the ratios of the ingredients used, the powder can be made to be faster or slower, and stronger or weaker, and can thus be tailored for a specific application.

The three ingredients used in gunpowder are charcoal, sulfur, and saltpetre. A fourth necessary component is water, though it is not usually listed as an ingredient.

Charcoal:

Charcoal:

Many types of charcoal can be used to make gunpowder, though a soft wood such as alder or willow is usually considered to be the best. The general wisdom is that, when the charcoal is milled with the other ingredients, a highly-porous wood allows the sulfur and saltpetre to be forced into the charcoal, thus producing a closer union of the ingredients.

There are two basic ways to manufacture charcoal. The method used in the 18th century was to stack wood in either pits or mounds. The wood was then covered with clay or mud with a small hole left for smoke to escape. The wood was set afire, and burned at a controlled rate with little oxygen until it was fully carbonized. This process could take weeks, and required some skill to keep the wood burning at the correct rate and to judge when the process was complete. For resinous woods such as pine, a pipe can be placed at the bottom of the pit or pile and used to draw off byproducts such as wood tar and turpentine.

The second method for making charcoal is the retort method, and is the principal process used to manufacture charcoal today. Wood is loaded into a container that is closed except for a small air hole. The container is then heated until the wood has turned into charcoal. In addition to collecting the liquid byproducts, additional compounds can also be recovered from the exhaust gases. The first industrial use of the retort method is claimed to have been started in 1786 by Bishop *Richard Watson* of England.

Both methods of charcoal manufacture, when conducted by a skilled operator, will produce a high quality product. The charcoal works equally well for gunpowder regardless of the method. Some, notably Bishop *Watson* and Sir *William Congreve*, have contended that the charcoal produced by the retort

method

method is superior to that produced by the pit method. However, period works on making gunpowder suggest that the gunpowder manufacturer can select the best available charcoal, and that this is of very high quality. Furthermore, once the retort technique became known, many gunpowder manufacturers did not switch to the new style of charcoal, but continued to buy charcoal from trusted producers using the older method.

Sulfur:

Sulfur, also called brimstone, is often found in a nearly pure state in the vicinity of volcanoes. In the 18th century, this was the principal source of sulfur, though it was also refined from various pyrites and other ores. If the sulfur is not of sufficient purity, it is easily refined by vaporizing it and letting the vapor condense on a cool metal plate. Most of the sulfur used in Europe and America in the 18th century was mined and processed in Italy.

Saltpetre:

Saltpetre, also called niter or potassium nitrate, is the least consistent of the ingredients used in gunpowder. In the 18th century, some nations, such as England and the Netherlands, imported most of their saltpetre from India. This saltpetre was refined in India before shipment, and then reprocessed to further refine it once it reached its destination country. Some nations, such as France, produced much of their own saltpetre.

The process used to manufacture saltpetre in the 18th century was to leach it out of soil and then refine it. It typically took two to three bushels of soil to yield one pound of saltpetre. The most common soil used was the earth that forms the floors of horse, sheep, or cow stables, though common cellar dirt was frequently used. The soil is soaked in water for a period of ten or twelve

hours,

hours, then the water is drawn off. The water now contains the dissolved saltpetre as well as many impurities and is called lye.

This lye-water is boiled to reduce its volume, then filtered through straw and wood ashes. The straw acts as a simple mechanical filter to remove soil particles. Wood ash contains potassium carbonate (potash), and this combines with calcium salts in the lye and precipitates out as calcium carbonate.

The next step is to separate the saltpetre from other salts. In addition to the saltpetre, the lye-water frequently contains common salt (sodium chloride) and sal ammoniac (ammonium chloride). If the lye was left in contact with the ashes too long during the filtering, then it may also contain potash. If an insufficient quantity of ashes was used, then the lye may contain muriate of lime (calcium chloride) and calcium carbonate. The separation of salts is done by reducing the lye-water by boiling until it is dense enough, removing any film or grease on the top of the lye, and then skimming the remaining upper portion of the lye-water to exclude the common salt which will have sunk to the bottom. This skimmed substance is boiled to reduce it further, then left to cool and form into crystals. After much of the saltpetre has crystallized, any remaining liquid is discarded to exclude the more soluble salts, such as sal ammoniac.

The refinement of the saltpetre was imprecise in the 18th century. Saltpetre of dubious purity would be refined again, but was still frequently left associated with other salts. When saltpetre was imported from India, the British government stipulated that it should have no more than 15 percent impurities. France purchased its saltpetre from local producers, and required no more than 30 percent impurities. In both nations, the saltpetre was refined two or three times further

further to improve it enough to make gunpowder.

Gunpowder made for the British government required double-refined saltpetre. Sporting powder for private trade and for hunting was made with triple-refined saltpetre. Powder used for blasting and for the Africa trade was made with only single-refined saltpetre.

In 1775, France offered a prize for improvement of the manufacture of saltpetre, with papers on this topic accepted until 1782. *Lavoisier*, the famous French chemist, was both a contributor and chairman of the commission which examined the papers. The result of this competition was a significant improvement in the production efficiency and the purity of saltpetre.

Manufacture:

Once the three ingredients were available, gunpowder could be manufactured. The common English powder of the 18th century used 15 parts saltpetre, 3 parts charcoal, and 2 parts sulfur. These are the same proportions used in most gunpowder today. The French gunpowder of the 17th and 18th centuries used 6 parts saltpetre and 1 part each of charcoal and sulfur.

The charcoal and sulfur are ground together until they are very fine, like talc. The saltpetre is ground separately until it also is very fine. The charcoal-sulfur mixture is combined with the saltpetre and a small quantity of water to form a paste. This paste is ground for anywhere from five to twenty-four hours, keeping the powder moist at all times lest it explode. The grinding can be done either in a wheel mill, using stone mill wheels much like those used in a grist mill, or in a stamp mill, by pounding it with wooden pestles in a mortar. For small scale operations, grinding is sometimes done in a ball mill, where the material is tumbled in a drum that also contains lead balls.

The

The goal of milling the powder is to incorporate the three components together by having as small particles as possible of each of the materials in close proximity to one another. The duration of the milling affects the incorporation, with longer times producing better results. English government contracts stipulated a minimum of five hours, while the French government used a minimum of 12 hours. Incorporation is the most dangerous part of powder making, and even in professional manufacturing operations, explosions occurred and workers were killed. After grinding, the powder is called milled powder or mealed powder.

If the powder was incorporated in a wheel mill, then the result tended to be a very fine low-density powder. To produce a more consistent result, the powder would then be pressed in a screw or hydraulic press. A stack of alternating copper sheets and inch-thick layers of powder was placed in the press, and then compacted into blocks over the course of half an hour or more. By pressing the powder, the density and therefore the quality becomes more uniform. The pressed powder becomes more robust in transport, being less likely to crumble and break apart. Also, pressed powder does not absorb water as readily as unpressed powder. After pressing, the powder blocks are removed from the copper sheets and then crushed and grained as with unpressed powder. Some British powder in the 18th century may have been made with wheel mills and not pressed.

When powder was incorporated using a stamp mill, pressing wasn't performed, as the stamping action tended to press it adequately, though not as consistently as when a hydraulic press was employed.

After milling and pressing as needed, the powder is grained. This process is also called corning. In the 18th century, this was done either by hand or by machine.

To

To corn the powder by hand, the slightly damp powder was rubbed across a screen with holes of a particular size, forcing the powder to extrude through the holes as short lengths with a specific diameter. To corn the powder by machine, the powder was placed on a vibrating screen that also contained a number of wooden blocks. The blocks would knock together to break up the powder, which would then fall through the screen once the particles became small enough.

In either case, the screens were made of leather with holes punched into it. The size of the holes determined the fineness of the powder. The resulting grains were sifted to remove the fine particles and dust. The fines and dust were reprocessed with the next batch of milled powder.

After graining, the powder can be glazed. When purchasing commercial British gunpowder, either dull (unglazed) or bright (glazed) powder could be requested. Glazing was not used at all in Britain prior to 1680, and was not required for British government gunpowder in the 18th century. It is likely that glazing was commonly done in France.

The powder was glazed by placing it in a revolving drum for several hours to smooth and polish it. Polishing the powder makes it more robust in transport, as it is less likely to break apart into dust. After glazing, the powder was once again sifted to remove dust.

A small quantity of graphite can be added during the polishing process to coat the powder. Glazing the powder with graphite reduces the ignition speed and also makes the powder less susceptible to moisture. Since the ignition speed is reduced, a smaller grain size of coated powder will have similar explosive properties to a larger unglazed grain size. In the 18th century, glazed powder referred to any polished powder. In

modern

modern parlance, glazed powder generally means that it has been coated with graphite.

Lastly, the powder was dried and once more sifted to remove the fines and to sort it into final grain size divisions. Drying was done in special rooms fitted with stoves.

Comparison of Gunpowder:

The principal differences between the gunpowder of *Vauban*'s time, that of *Muller*'s time, and the present are the quality of ingredients used in the powder and the repeatability of the manufacturing process. Variability in both the ingredients and the manufacturing contribute to inconsistency in the explosive force and in the combustion speed of the powder. Both cause differences in the density of the powder. Various methods used in the testing of gunpowder also resulted in powders with different characteristics.

The quality of the saltpetre used in gunpowder varied quite considerably based on how it was refined. Major improvements in saltpetre refinement occurred in the 1770s and 1780s under the direction of the French chemist *Lavoisier*. He created a set of reliable tests to check the purity of the saltpetre and ensure that it was free of foreign salts and other matter. The quality of French gunpowder improved, and that of other nations gradually followed.

The quality of charcoal also varied. Although high-quality charcoal could be selected from charcoal made using the pit method, the process relied heavily on an experienced manufacturer. In 1786, the retort method of charcoal was first used commercially. This produced consistent high-quality charcoal which required less skill from the maker. The new retort charcoal gradually replaced the older pit method in England. Other nations were slower to follow.

The

The exact ratio of ingredients used in gunpowder varied from batch to batch. In part, this was due to the initial moisture content of the ingredients. Although the ingredients were measured by weight, they were not dried consistently beforehand, and the powder made in rainy weather was distinctly different from the powder made in dry weather. *Congreve* demonstrated this problem in the 1790s, and, by carefully drying each of the initial ingredients, improved the consistency between batches of powder.

Incorporation of the ingredients varied from batch to batch based on the milling process. Stamp mills produced reliable, consistent results, but took a long time to perform their work. Wheel mills were used because they saved time and were safer, but initially they did not produce the same quality of results. In the 19th century, the wheel mill was perfected under the oversight of the American gunpowder manufacturer *Du Pont*, at which point the wheel mill produced more consistent results than the venerable stamp mill.

The density of the gunpowder varied from batch to batch. Some of the this variation is caused by the size of the granulation of the powder. The exact composition and the moisture content of the powder also affect the density. Eighteenth-century sources list powder densities from as low as 50 $\frac{3}{4}$ pounds per cubic foot to as much as 95 $\frac{1}{4}$ pounds per cubic foot. *Muller*'s figure is 55 pounds per cubic foot. A modern measurement of uncompressed Goex brand FFFg sporting powder gives a value of 59.4 pounds per cubic foot, while the slightly coarser FFg grade is 60.0 pounds per cubic foot. By consistently pressing the powder, the density variation has been virtually eliminated, but this was not fully done until the early 19th century.

Characteristics

Characteristics of Powder:

Different ratios of ingredients were used purposefully to affect the speed at which the gunpowder combusts, with a lower saltpetre content producing a slower powder. The size of the grains of gunpowder also affect its speed. Gunpowder combusts on the surface of the grains, so for the same weight of powder, smaller grains burn faster.

For the gunpowder to be effective in firing a gun, it must combust while the projectile is still within the barrel of the weapon. However, if it combusts too fast it will cause undue strain on the gun barrel. For small arms such as pistols and muskets, a fast powder is required because of the small length and volume of the gun barrel. For cannons, a slower powder is preferable, as the projectile is then accelerated for the entire length of the barrel. For a given speed of powder, a higher quality powder will produce a faster muzzle velocity, because the combustion of the gunpowder produces a greater quantity of gas, which exerts a greater pressure on the projectile.

Vauban wrote his memoir on the defence of fortifications around 1667. At that time, only one or perhaps two grades of gunpowder were manufactured for the military, and it is probable that a fort would have only a single grade. The preponderant use of powder was for cannons and mines, both of which work best with a comparatively slow powder. It is probable that this slow powder was also used for muskets. In addition, the French government's test for the quality of gunpowder was to fire a ball from a standard-sized mortar and ensure that it travelled a certain minimum distance; this helped optimize the powder for artillery rather than for small arms.

In *Muller*'s time, the English government requested only a single grade of gunpowder for military use. This

was

was a comparatively fast powder, better suited to muskets and pistols than to cannons and mines. The English method of proving the quality of the gunpowder was to fire a sample of the powder from a pistol-like eprouvette and then measure the distance a spring moved. Using a short-barrelled test helped to optimize the powder for small arms.

Commercially, in the 18th century, different grades of powder were available. Some of these had different grain sizes, with four sizes being typical. These sizes correspond directly to modern powder, being designated C or cannon grade, F or musket grade, FF or rifle grade, and FFF or pistol grade. Lower saltpetre powders were also sold, mainly for industrial mining ore and for trade with Africa. The powder sold to the military in both France and England was predominantly a single grade made to the highest quality standards of the time.

Amount of Gunpowder Required

Muller's calculation for the amount of powder required for different purposes during a siege, as detailed on pages 135 through 137, is largely copied from *Vauban*'s writings. However, to someone experienced with using modern black powder, the amount allowed for firing cannons and muskets will appear excessive.

For example, *Vauban* gives a load of between 189 and 252 grains per charge for a musket of approximately 0.69 caliber. This corresponds to the French Charleville muskets in use at that time. During the late 1750s, the British government allotted a half ounce of powder, or 219 grains, per round in their muskets, which were predominantly 0.75 caliber. Today, when target shooting using modern powder using a musket of 0.69 caliber, a powder load of around 90 grains is commonly used.

used. Even with this charge, one can see burning powder being flung from the muzzle of the musket as it fires.

If *Vauban*'s estimates for the amount of gunpowder required for a siege are correct, then *Muller*'s values are almost certainly over-estimated. Since the speed of English powder was greater, less was required for small arms. Since the quality of the powder had improved, less was required for cannons and mines.

Since *Muller*'s time additional improvements have been made in the quality and uniformity of gunpowder, and thus *Muller*'s estimates are most likely overstated for modern powder. Modern estimates of powder usage take into account only the amount used for individual loads. Both *Vauban* and *Muller* are including some amount to cover losses due to spillage and spoilage, and are also probably adding a safety factor in case of a prolonged siege or campaign.

The best batches of gunpowder manufactured in the 18th century were likely of equal strength to the powder of today, but the average military powder was substantially weaker due to lower purity of ingredients and poorer manufacturing. Some sources cite that the work of *Congreve* and *Lavoisier* reduced the necessary powder charge for a musket or cannon by one-third, though this seems to include all of the improvements from *Vauban*'s time.

READING MAPS

THE modern reader is probably much more familiar with topographic maps than with the style of maps presented in 18th century books. Whereas on a topographic map, there are lines indicating areas with exactly the same elevation, on period maps only

significant

significant features are marked.

On period maps, steep areas are drawn with hatched lines. The closer the lines are to one another, the steeper the area. Man-made features, such as the glacis of a fortification, are also drawn with a great many lines. Rather than using color or shading to show terrain type, small trees or tufts of grass are shown.

See Plate N-I and Plate N-II for a comparison of a map from 1777 and a relatively modern 1:24,000 scale United States Geographic Survey map from 1950. The two maps are the same orientation and scale, and show the region of Fort Ticonderoga on Lake Champlain in the State of New York. This fort is currently maintained as a privately owned historic site, and thus it is in much the same configuration today as it was in 1777. In addition to the fort itself, which consists of four bastions and two ravelins, there are a number of redouts, a line with tenailles and redans, and a star fort across the lake.

Some of the differences between the maps are due to intentional modification of the landscape, such as the addition of roads and the removal of a floating bridge. Other differences are inaccuracies in the original survey or source map used. One significant feature that differs is the place marked Sugar Loaf Hill to the southwest of Fort Ticonderoga on the 1777 map. This is probably the same hill that is marked Mount Defiance on the modern map.

There is an approximately 5 degree difference in where true north is marked on the two maps. True north is shown by a stylized arrow at the top of the 1777 map. On the modern map, grid north (approximately the same as true north) is in the orientation of the solid line near the right side of the map.

Plate N-III shows a detail of the 1777 map. The fort is more easily visible, as are a number of the redouts and the

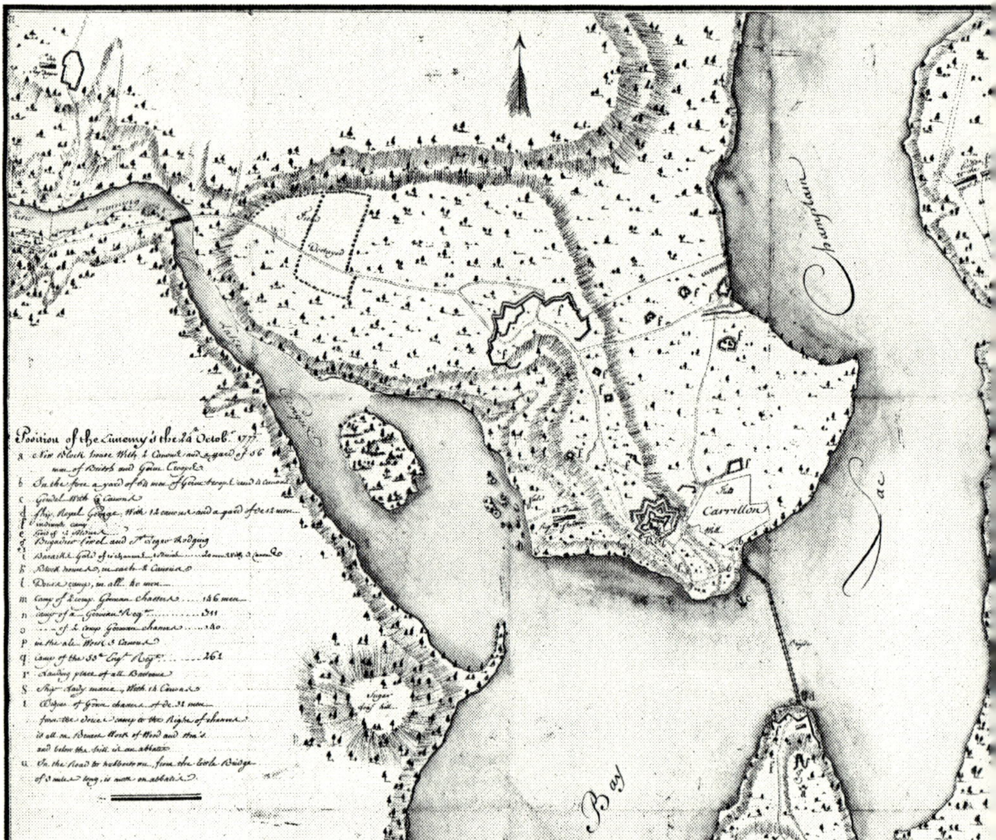

Environs of Fort Ticonderoga, 1777

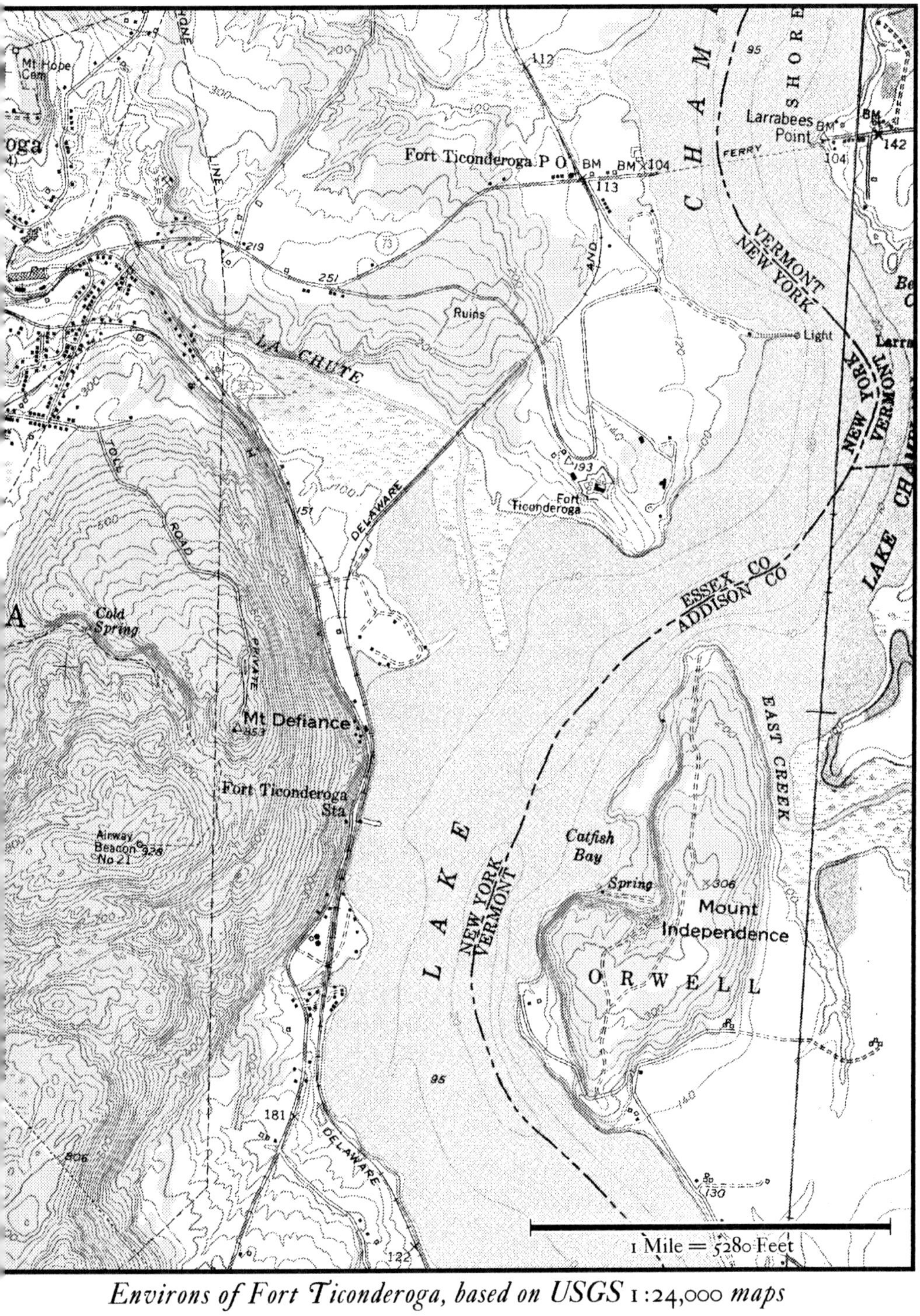

Environs of Fort Ticonderoga, based on USGS 1:24,000 *maps*

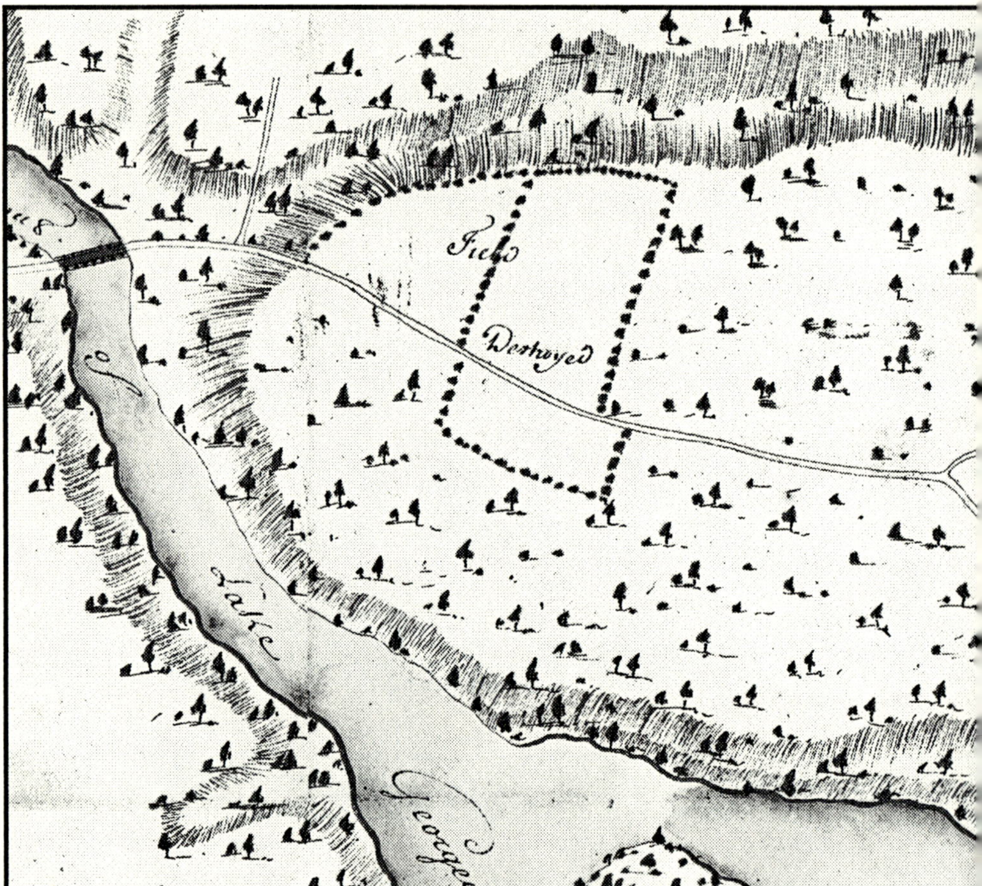

Detail from Map of

Fort Ticonderoga, 1777

the defensive line. The terrain variations are clearly shown as hatch marks, and the ground cover is shown by small drawings of trees and scrub grasses.

This map is from the American Revolutionary War and represents the state of fortifications and troops shortly after the English captured Fort Ticonderoga without conducting a siege, and with relatively minimal resistance from the Americans. They accomplished this by bringing cannons to the top of Mount Defiance, and thus commanding the fort. Ironically, it is likely that some of the American engineers carried copies of *Muller*'s books with them, and knew that Mount Defiance needed to be defended. It is certain that the American commander, Major General *Arthur St. Clair*, knew that this defense was necessary, but he had taken command too late and the previous commander of the fort, General *Horatio Gates*, was derelict in securing this point, feeling that it would be impossible to bring cannons up the hill.

A LEXICON *of* FORTIFICATION

ACCELERATING A SIEGE: If the defending garrison is very weak or the assaulting force is very numerous, then the siege may be shortened by opening the trenches closer to the Place or by not constructing the lines of circumvallation. See page 127.

AIM FRONTLET: Also called a Dispart or a Dispart Frontlet. A piece of wood fitted to the muzzle of a cannon, tall enough to match the height of the breech. The wood has a notch for sighting at the top, and allows the cannon to be directly sighted for straight shots. It was seldom used, as almost all cannon shots were significantly arced. See page 41 and Plate I.

ANGLE: See Salient Angle or Re-entring Angle.

APPROACH:

APPROACH: The approaches are all of the trenches, saps, and other works that are carried toward the Place that is besieged, including the parallels, epaulements, redouts, places of arms, mine galleries, and lodgments. The term is sometimes used more specifically to only indicate the trenches which zig-zag towards the Place.

À RICOCHET: See Ricochet.

ARROW: A work placed at the salient angles of the glacis, consisting of two parapets, each 80 yards long. This work has a ditch in front of it of 10 to 12 yards across, and is connected to the covert-way via a caponiere that is 24 to 30 feet wide. See Plate N-VII.

ARTILLERY: Also called Great Arms. Cannons, howitzes, and mortars.

ARTILLERY-PARK: A place in the camp where all of the artillery not currently in service are stationed. Generally, the cannons, howitzes, and mortars are lined up by regiment in ordered rows, with the ammunition wagons directly behind the appropriate guns. The artillery pieces are separated by enough space that any regiment may move without the remaining regiments being in the way. See page 17.

ASSAULT: Also called Insult. A sudden furious attack on a Place, typically through a breach, where the attackers are not covered by a work. While an assault is being made, the batteries cease firing at that location so as to not to cause mischief to their own troops.

ATTACK: Taking a Place without a siege, or storming out of the front of the trenches or galleries to take the Place. Alternately, the set of approaches and course of actions made by an army to get possession of a fortified place.

AUGET: A wooden trough placed in the floor of a gallery to connect a mine with the entrance of the gallery. This is used to prevent crushing the saucisson when the gallery is filled. See page 219.

AUGRE: Also spelled auger. A long metal bar with a flat blade used to probe the ground to determine if there are any mines. This is often used by miners to detect counter-mines. The blade of the augre is wider than the bar to allow a hole to be bored in the earth without the shaft of the augre getting stuck.

BALL: The projectile used in muskets, wall-pieces, and cannons. Balls for muskets and wall-pieces are made of lead, while balls for cannons are made of iron. An iron cannon ball of 4 inches in diameter weighs 9 pounds, and a lead ball of 0.662 inches in diameter weighs 1 ounce. As there were often different calibers of both cannons and muskets, either every ball size would be stockpiled for a siege, or the ability to cast the balls as required would be needed.

BANQUETTE: Also spelled Banquet. A raised platform of earth or wooden boards on the inside of a parapet. Soldiers stand on the banquette to fire their muskets over the top of the parapet. See Plate N-IV and Plate N-V.

BARBET: Guns are firing *en barbet* when then can shoot over the parapet of a work without requiring embrasures. In this case, the parapet or breastwork must be no more than 3 feet high. This is more expeditious, but less safe, than firing through embrasures. A garrison can usually fire many guns *en barbet* until the attackers have nearly completed their batteries.

BASKET: Baskets are used for a variety of purposes, mainly to transport earth. Baskets used for carrying soil and rocks must be made sturdy enough that they will withstand rough handling, and should have a reinforced bottom. Baskets also are a type of tapered gabion used for defense along a parapet. In this case, they are 18 inches high, have no bottom, and range in diameter from 18 inches at the top to 8 inches at the base. They are placed touching each other on the parapet and are filled with earth to form a sort of embrasure for the soldiers.

BASTION: A part of the main fortification forming one of the points of the fort. The bastion consists of two outward facing sides, two flanks which abut the walls of the work, and an opening into the main body of the work called the gorge. A bastion is called empty if the ground inside of it is much lower than its parapets. A bastion is called full if the ground inside is even with the rampart, with only the parapets higher than the rest. A tower bastion has a building higher than the parapets in its interior. See Plate N-V, Plate N-VI, and Plate N-VII.

BATTERING IN BREACH: The act of firing the batteries so as to create a hole in a wall of the work so that it can be assaulted. The breach is started at the bottom of the wall so that the rubble does not cover and further protect the wall. The alternative to battering in breach is to use mines to create a breach. See page 60 and Plate X.

BATTERY, CANNON: Any place where cannons are mounted, either to attack the enemy or batter in breach. The battery is constructed with an epaulement, 8 feet high and 18 to 20 feet thick, with embrasures made in it to allow the guns to fire. When a battery is erected for battering in breach, it is often sunken in the glacis so as to allow the cannons to bear on the wall being battered. Platforms are made for the guns to prevent their recoil from damaging the works and to make it easier to service the guns. The term battery both refers to the work that protects the guns and the guns themselves. See page 35 and Plate VII.

BATTERY, MORTAR: Any place where mortars are mounted to attack the enemy. They consist of a epaulement of 18 to 20 feet thick with a height of 7 ½ feet on the side next to the mortars and a height of 6 feet

feet on the side facing the work. There are no embrasures. If the mortars have sufficiently sturdy beds, no platforms are necessary. See page 42 and Plate VIII.

BAYONET: A metal blade that is affixed to the muzzle end of a musket to covert the weapon into something akin to a pike. Although it had great use in deciding field battles, in sieges it was typically only used as a last resort by a garrison that was running low on powder and was repelling an attack. The invention of the bayonet is often credited to *Vauban*.

BLIND: A wooden frame used to shore up walls in trenches and saps, and the walls and roofs of galleries. See page 4 and Plate I.

BLOCKADE: To completely surround a Place such that no provisions can possibly enter, forcing the Place to eventually capitulate for want of food. Blockades rarely work except when a Place is known to be low on provisions and no relieving army is likely to arrive.

BOAT: Aside from the usual meaning, simple boats can be constructed to carry one or two men at a time across a wet ditch. These boats can be made of folded tin or baskets covered in oil-cloth, and are not expected to be durable. See page 124.

BODY OF A PLACE: This can either mean the buildings within a fortification or the main fortification itself. It excludes outworks such as ravelins and lunettes.

BOMB: A shell fired from howitzes or mortars. This is a hollow sphere of iron that is filled with gunpowder and musket balls, pieces of scrap iron, nails, or other metal objects. It has a timed fuse that, when made the proper length, will cause the bomb to explode just before it strikes the ground.

BOMB-PROOF: An underground chamber that is excavated deep enough so that a barrage of shells cannot get through. These are often made within filled bastions.

bastions. See Plate N-VII.

BONNET: A small work placed before the salient angle of a ravelin to cover it. Typically, it consists of two parapets of three feet high and 10 to 12 feet broad that are joined in an angle identical to that of the ravelin. The bonnet is similar to a counter-guard, except that it is always before a ravelin, and doesn't cover the whole of the ravelin's faces, usually due to the presence of lunettes. See Plate N-VI.

BOOM: A floating barrier used to prevent boats and ships from getting into a harbor. This is often a series of logs or manufactured floats chained to one another. While it may not prevent a ship from crossing it, the boom slows the vessel down enough that guns can be brought to bear upon it.

BREACH: An opening made in a wall or rampart of the fortification either by battering or mining. It must be sufficiently wide for a body of troops to enter and either drive the besieged out of it or to force the capitulation of the garrison.

BREASTWORK: See Parapet.

BRIDGES OF COMMUNICATION: Bridges made over rivers, streams, ravines, and marshes to allow different parts of an army or two separate allied armies the ability to come to each others aid. This term is also applied to bridges between two fortifications that provide each other mutual support across a river.

CAISSON: Also called a Coffer. A wooden box, typically 3 or 4 feet square made with planks 1 or 2 inches thick. The box is loaded with from 4 to 20 charged shells, and loose powder is placed on top. The box is buried five or six feet deep. A fuse or saucisson is run to the box so that when the enemy have taken possession of the ground above it, the caisson can be ignited. The loose powder serves to blow up the ground above the caisson. The shells are timed to

explode

explode a few seconds later, in order to cause as much mischief amongst the enemy as possible.

CANISTER-SHOT: Also called case-shot. A type of ammunition for cannons where a number of small bullets or balls are enclosed in a tin cylinder. The cylinder ruptures when the gun is fired, and the shot is dispersed in a broad pattern.

CANNON: An artillery piece designed to fire at a low angle. Cannon are used against troops and for battering a fortification. The size of a cannon is given as the weight of a solid iron ball that can be fired from it. For example, the standard British 24 pound cannon has a bore of 5.55 inches and is 9 feet long.

CAPITAL: The capital of a work is the imaginary line that divides it into two equal and similar parts. For example, the capital of a bastion bisects the salient angle.

CAPITULATION: The agreement made between the besiegers and the besieged on the conditions of surrender. See page 197.

CAPONIERE: Also spelled Caponier. A passage made from one work to another. Caponieres are usually 10 to 12 feet wide and 5 feet deep, and covered on each side with a parapet. The far side of the parapet terminates in a slope or glacis. At times, the caponiere may be covered with planks and earth. For example, when the ditch is dry, caponieres connect the curtain to the ravelin, and connect the covert-way to the arrows. Caponieres also refer to single parapets at the entrance of a ditch before a ravelin, which are raised as needed during a siege, and used by infantry and light artillery to dispute passage over the ditch. See Plate N-IV, Plate N-V, and Plate N-VII.

CASEMAT: Also spelled Casemate, Cazemate, or Cazematte. Typically a vault or underground area in the flank of a bastion with embrasures and loop-holes to

allow

allow cannons and muskets to fire out of them to defend the opposite bastion. Casemat also refers to the lower level of a tower in a tower bastion which allows guns to be protected behind an additional work and further defend the bastion. See page 78, page 93, and Plate N-IV.

CASTLE: An antiquated fortification used to defend a strong point or small town. This refers to fortifications made before artillery was employed.

CAUSEWAY: A raised roadway crossing a marsh or low ground. Causeways are often built as bridges of communication by the besieging army, and may be required as part of the attack on the Place. See page 98.

CAVALIER: An earthwork raised to gain command of another work, used in both the attack and defence of a Place. When cavaliers are made during an attack, they are made of earth and gabions and located partway along the glacis with the intent of enfilading or exposing the covert-way. When a cavalier is used for defence, it is made in the body of the work and raised 10 to 12 feet higher than the work in order to command the surrounding works and ground. Cavaliers are usually located in the center of a bastion, though sometimes they are placed along a curtain or in the gorge of a bastion. They are seldom used except when the Place itself is commanded by a hill or other rising ground.

CAVALRY: Soldiers who serve and fight from horseback. These are divided into *horse* or *horse-guards*, who fight solely from horseback, *dragoons*, who can fight either on foot or from horseback, and *hussars* or *light-horse*, whose principal advantage is their speed and maneuverability. The horse-guards, being the heaviest equipped, are used to push an enemy from an unfortified field. The light-horse are used for reconnaissance and quick flanking actions. The dragoons are used against infantry and in wooded terrain.

CHAIN:

CHAIN: A chain stretched across a river or harbor mouth to prevent the passage of ships. The chain frequently has floats attached to it to keep it near the surface of the water.

CHAMADE: A rhythm beat on the drum to signal a desire to parley. This is done when the garrison wishes to surrender, and also for a cease-fire to allow both sides to bury their dead. See page 197.

CHAMBER: The place in a mine where the powder is lodged. This is the focus of the mine and is at the mathematical focus of the paraboloid of earth that is blown up when the mine is sprung.

CHANDELIER: A frame used to hold fascines as a temporary screen to protect workmen from enemy fire. See page 4 and Plate I.

CHAUSS-TRAP: See Crows-Foot.

CHEVAUX DE FRIZE: A thick beam with a large number of crossing wooden stakes at right angles to the beam and each other. The beam is often 10 to 12 feet in length, and the stakes are 6 feet long and 1 $\frac{1}{2}$ inches in diameter. The stakes are either pointed or tipped with nails. Chevaux de frize are placed in breaches and in gateways to impede horse and foot movement. See page 5 and Plate I.

CIRCUMVALLATION: A fortification consisting of a parapet and a ditch that the besiegers make to protect themselves from forces that may try to relieve the garrison. The line of circumvallation is placed out of cannon shot of the camp to prevent a relieving army from assailing the besiegers with artillery. The line of circumvallation is typically made 12 feet wide and 7 feet deep, with a parapet on top of it, and periodically strengthened with redouts. See page 12 and Plate II.

CITADEL: A fort with four to six bastions in the center of a city or other fortified place. The citadel may also be half within and half without the walls of the

Place.

Place. The difference between a citadel and a castle is that a citadel has bastions.

COFFER: See Caisson.

COMMAND: A hill or rising ground commands a fortification when a cannon placed on the hill can be fired into the fortification, especially when the hill has sufficient elevation that the fort cannot respond in turn.

COUNTER-APPROACH: The counter-approaches are all of the trenches, mines, and other works that are carried by the besieged toward the besiegers in order to better defend the Place against a specific attack.

COUNTERFORT: Pillars or buttresses spaced 15 to 20 feet apart along the inside of a wall to strengthen both the wall and rampart. Counterforts are built in many earthen walls, but are usually hidden from view by the dirt and turf. When a wall has a revetement, the counterforts are directly attached to the revetement. When there is a casemat, the counterforts are often visible as buttresses within the casemat.

COUNTER-GUARD: Small ramparts with a parapet and ditch to cover the faces of a bastion or ravelin from breeching fire. They consist of two faces parallel to the faces of the bastion or ravelin, and joined at a salient angle. The counter-guard covers all of both faces of a ravelin, or almost the entirety of both faces of a bastion, and is located between the ravelin or bastion and the covert-way. See Plate N-VI.

COUNTER-MINE: Mines made by the besieged either to counteract the mines of the besiegers or to disrupt their attacks. Counter-mines are made both when the Place is originally fortified and also once the Place is under attack.

COUNTERSCARP: The wall of the ditch on the side away from the fortified place. This applies to any ditch, but is often used more specifically to refer to the location between the covert-way and the main ditch.

See

See Plate N-IV and Plate N-V.

COUNTERVALLATION: A fortification consisting of a parapet and a ditch that the besiegers make to protect themselves from sallies and assaults from the besieged garrison. See page 18.

COVERT-WAY: Also spelled Covered-way. A flat area 30 to 36 feet wide that goes entirely around the works of the fortification, excepting where there are cliffs and rivers. The covert-way is between the main ditch of the Place and the glacis, and has a parapet of 7½ feet in height on the glacis side. The counterscarp separates the covert-way from the main ditch. See Plate N-IV and Plate N-V.

CROWN-WORK: A detached work that occupies a great amount of ground and that is placed in front of a bastion or a curtain. It consists of two sides running away from the main Place connected with two half bastions and one full bastion, with a large gorge on the side of the Place. Crown-works are generally made to enclose a building that cannot be relocated within the fortified place or to hold rising ground. Note that Plate XIII and Plate XIX show horn-works, not crown-works. The two differ in size, with the crown-work having an additional bastion between the outermost points of the crown. See page 79 and Plate N-VII.

CROWS-FOOT: Also called a Chauss-Trap or Caltrop. A four-pointed iron made such that no matter how it falls, one point is always uppermost. Crows-feet range from 4 to 7 inches long, with the short ones used on hard surfaces and the long ones used in soft earth. Their purpose is to impede cavalry by endangering the feet of the horses. See page 5 and Plate I.

CURTAIN: The generally straight wall that joins the flank of one bastion to the flank of the next. See Plate N-V.

DEMI-BASTION: Also called a Shoulder. This is

when

when an otherwise straight wall, such as the flank of a horn-work or crown-work, has a step in it, creating an extra face so as to cover the portion of the wall between the shoulder and the point of the work. See Plate N-VI and Plate N-VII.

DEMILUNE: See Ravelin.

DESCENT: A passage beginning on the glacis, travelling under the covert-way, and opening into the bottom of the ditch. For wet ditches, the descent opens onto the level of the water. At the beginning, the descent has the form of a trench covered in blinds, fascines, and possibly gabions. If the ditch is deep enough, the descent takes the form of a sap or gallery. See Plate XI and Plate XII.

DETACHED BASTION: A ravelin with flanks. This is much like a bastion, still located within the covert-way, but separate from the main works. This term is also used to refer to a counterguard with flanks. See Plate N-VI.

DETACHED REDOUT: A work made outside of the covert-way. This is a ravelin with flanks, identical to a detached bastion except for its location. Detached redouts are used to force the enemy to start their trenches further from the main body of the work, and should be located within musket shot of the covert-way so as to be protected from there. See Plate N-VII.

DISPART FRONTLET: See Aim Frontlet.

DIRECTOR: The head engineer in charge of the attack.

DITCH: A large deep trench made around the entire work. The earth dug out of it is used to raise the ramparts and parapets. The width of the ditch varies, but 60 to 96 feet is typical. Ditches can be left dry, kept wet, or designed so that a sluice can be used to fill or drain them at pleasure. For wet ditches, the mud is dredged from the bottom periodically; otherwise the

difficulty

difficulty of the passage over the ditch will be much reduced. See Plate N-IV and Plate N-V.

DOUBLE DITCH: A ditch with a deeper ditch running down the middle of it. The purpose is to force the attackers to make two separate descents. See Plate N-IV.

DOUBLE-SAP: A sap made twice as wide as normal. Two groups of sappers advance the sap, each group pushing a stuffed gabion in front of them. Periodically, the sap has traverses so that firing à ricochet has little effect. These traverses have passages around each side, rather than around just one as in an ordinary single sap. See page 51 and Plate V.

DRAGOON: A soldier who serves as mounted infantry. Dragoons can fight from horseback in the manner of the regular cavalry, and can also fight on foot.

EARTHWORK: Any fortification made principally from piled soil. The sides of an earthwork may be reinforced with palisades.

EMBRASURE: An opening made in the flanks of a fortification or the parapets of a battery. These are typically about 2 ½ feet wide on the inside of the fortification, 8 to 9 feet wide on the outside, and 3 feet above the interior ground. They are made to allow cannon to fire through a parapet. See Plate N-IV and Plate N-V.

EN BARBET: See Barbet.

ENFILADE: A work is enfiladed when a cannon or howitze can fire such that a shot will travel down the length of the parapet or along the length of the trench.

EPAULEMENT: A high breastwork, sometimes made with the help of filled gabions and fascines. In a siege, the besiegers generally raise an epaulement of 8 to 10 feet in height near the entrance to the approaches to screen the cavalry, which is placed there to support the guard of the trenches. An epaulement often refers to a high breastwork thrown up to defend the flank of a

work.

work. The epaulment serves as a detached orillon, and makes it harder to dismount cannon from the flank. An epaulment differs from a parapet in that it has no banquette or other means for the defenders to attack from it.

ESCALADE: To take a Place by a secret assault where ladders are used to pass the ditch and to climb the ramparts. Generally, a group of soldiers is also stationed on the counterscarp so that if the garrison raises the alarm, the defenders may be delayed in mounting the rampart and opposing the escalading force. See page 123.

ESCARP: Also called an Escarpment or a Scarp. The outside of the rampart next to the ditch. This is the wall of the ditch that is closest to the fortification. This term can also apply to anything high and steep. See Plate N-IV and Plate N-V.

EXCAVATION: The cavity or crater left after a mine has exploded.

FACE: The part of a work facing away from the main body of the fortified place. See Plate N-V.

FASCINE: A bundle of branches or sticks used to support earthworks and hinder enemy musket fire. The most common size used in trenches is 6 feet long and 6 to 8 inches in diameter. A smaller size, used in covered works, are sap-fagots, which are the same diameter but only 3 feet long. A larger size, used to reinforce the parapets of batteries, are saucissions, which are 10 to 16 feet long and 12 to 18 inches in diameter. Fascines are made by setting three pairs of pickets or stakes into the ground, with each pair crossing to form an X. The branches used for the fascines are piled in the crossed pickets, then bundled by tying them with willow or vines every 2 to 3 feet. Six men working, with 2 men cutting branches, 2 gathering them, and 2 binding them, can make 12 regular fascines an hour. See page 3 and

Plate

Plate I.

FATHOM: 6 feet. See the Notes on units of distance on page 254.

FAUSS-BRAY: Also spelled Fausse-bray. A low rampart going entirely around the work, raised about 3 feet above the rest of the ground, and with a parapet from 18 to 24 feet in front of the work. They are seldom used except in front of curtains in works designed by *Vauban*, and there they are more properly called tenailles.

FIELD-PIECE: A light cannon used to support troops in the field. Generally, a field-piece is no heavier than a six pound cannon, though 12 pound cannons may be used. A light field-piece refers to a two or three pound cannon.

FIRE-BALL: Also called a Light Ball. A projectile, launched from a mortar, designed to give off light and set wooden objects on fire. The general method of construction is to make a thick brown paper shell of the appropriate size for the mortar that will launch it. The shell is filled with a mixture of 3 parts gunpowder, 2 parts saltpetre, and 1 part sulfur, with a small quantity of rosin, iron filings, sawdust, and charcoal. The completed shell is covered in a mixture of melted pitch and rosin. Various small tubes with musket balls and powder charges can be added to the shell, causing musket balls to be fired randomly as the shell burns. When the intent is to provide light, the shell is filled with equal parts of sulfur, pitch, rosin, and gunpowder.

FIRE-WORK: Any item made to burn, give off light or smoke, and confuse the enemy. Fire-works consist of fire-balls, smoke-balls, and stink-balls. Smoke-balls are similar to fire-balls except that they are made with more pitch, rosin, and sawdust; when properly made, they smoke for half an hour. Stink-balls have the addition of ground horse hooves, asafœtida (resin from the roots of

wild

wild fennel), and other stinking herbs.

FLANK: In fortification, the flank is the part that joins the face of a bastion to the curtain. It is used to defend the curtain, the face of the opposite bastion, and the ditch along the curtain. The flanks are also used to oppose batteries on the counterscarp and glacis. When referring to troops, the flank is either side of the main force, and to flank an army means to fire upon the flank of the army from the side. See Plate N-V.

FLOAT: A sort of floating platform used by the besieged to remove troops from a detached work, such as a ravelin, across a wet ditch. See page 159.

FORK: See Sap-fork.

FORT: A fortification that does not enclose a town or city.

FOUGASS: A small mine, 6 to 8 feet underground and placed under the glacis or under a dry ditch. See page 214.

FRAISE: Also spelled Frize. Stakes or palisades placed horizontally outward on the outer slope of a rampart. These are placed to prevent the work from being taken by surprise. If the fraises are sloped slightly downward, it helps make grenades roll off into the ditch instead of damaging the works. Fraises are generally 7 to 8 feet in length and 5 inches in diameter. See Plate N-IV.

FRONTLET: See Aim Frontlet.

GABION: A cylindrical basket open at both ends. Gabions are often made of willow or osier. The gabions used for saps are 3 to 4 feet in height and 2 ½ to 3 feet in diameter. Those used for field fortifications, including batteries, range from 3 to 6 feet in height. Small gabions are sometimes used at the top of parapets to make a kind of embrasure for muskets; see Basket. Gabions are always placed empty and then filled with soil, except for stuffed gabions. Stuffed gabions are 4 to 6 feet

6 feet in height, and packed very full of branches and small pieces of wood. They are turned on their side and rolled before the first workman in a trench to protect against musket fire. See page 3 and Plate I.

GALLERY: A passage made underground leading to one or more mines. The standard dimensions are 4 feet wide and 4 ½ to 5 feet high. The earth above is supported by wooden frames and boards.

GARRISON: A body of troops composed of horse and foot, placed in a fortification to guard it in time of peace and defend it in time of war.

GATE: A large entrance into a fortified place, usually covered by a ravelin or horn-work. A gate always leads to a principal avenue or street, and usually has doors that can be shut in time of attack. See Plate N-V.

GLACIS: A gently sloping ground running smoothly from the parapet of the covert-way to the field beyond the fortification. The width of the glacis is at least 120 feet. The glacis is made smooth and even, with no trees, bushes, or sunken points to provide cover for attackers, and sloped so that the Place can fire across any part of it. See Plate N-IV and Plate N-V.

GORGE: The part of a work that is next to the body of a Place. For works that are part of the main Place, such as a bastion, the gorge is between the flanks of the work and opens directly into the main body of the Place. For detached works, such as a ravelin, the gorge opens onto the ditch. In either case, the gorge has no rampart or parapet. See Plate N-V.

GRAPE-SHOT: A type of ammunition for cannons and howitzers where a number of small balls are bound together on a wooden base and around a wooden stick, then wrapped with a canvas covering to keep them in place. The round of grape-shot is loaded into a cannon with the wooden base against the powder charge.

When

When fired, the balls separate from the base and are dispersed against the target. The grape-shot round for a 24 pound cannon is typically 85 four-ounce balls. For a 1 ½ pound field piece, the grape-shot round consists of nine 1 ⅛ ounce balls.

GREAT ARM: Any artillery. Cannons, howitzes, and mortars.

GRENADE: Also spelled Granade. A hollow iron ball 2 ½ to 3 inches in diameter and filled with fine powder. The opening is filled with a wooden plug with a fuse running through it. Grenadier troops ignite and throw grenades amongst the enemy during an attack, where the iron shell shatters and causes much mischief.

GUNPOWDER: An explosive composition used to fire guns great and small, set off mines, and create fireworks. See the Notes on gunpowder on page 263.

HADLEY'S SEXTANT: An instrument for measuring angles, functionally identical to the modern sextant. A double mirror system is used to give the user a simultaneous view of the two objects between which the angle is measured. When used for navigation, the sextant is used to measure the angle between a star and the horizon. In sieges, the instruments is used on its side to measure the angle between two points of the fortification, or the angle between an accessible place and a point on the fortification. Using triangulation between the fortification and known points away from it, an exact plan of the fortification can be made without getting within the range of its guns.

HALF-MOON: See Ravelin.

HARROW: A wooden bar with a great many iron teeth, like a very wide rake. The harrow is an agricultural implement used for leveling plowed land, breaking up clods, and covering seeds. In a siege, harrows can be thrown in the breach to make the passage for the besiegers more difficult. In this case, the harrow works

better

better if formed like a chevaux de frize.

HOOK: See Sap-fork.

HORN-WORK: A detached work that occupies a large area of ground and that is usually placed in front of a gate in a curtain. It consists of two sides running away from the main Place connected with two half bastions and a short curtain. Horn-works are generally made to include a rising ground or to guard a principal avenue into a town. A horn-work is a smaller version of a crown-work. See page 79, Plate XIII, Plate XIX, and Plate N-VI.

HORSE: A general term for all sorts of cavalry.

HORSE-GUARD: Cavalry that fights solely from horseback. This is in contrast to dragoons, who are trained to also fight on foot.

HOWITZE: Also spelled Howitz and Howitzer. An artillery piece somewhere in design between a mortar and a cannon. These are used as mortars to lob shells into a Place, but generally throw their shells at a shallower angle, which allows the shells to be fired à ricochet. The physical difference between a howitze and a mortar is that the howitze has its trunnions in the middle of its barrel, while a mortar has its trunnions at the breech end of its barrel. Howitzes also tend to be longer than mortars for an equivalent bore size.

HURTER: A stout piece of wood about 9 feet long and 6 inches square that is placed against the parapet of a battery to prevent the cannon wheels from damaging the parapet. See page 40.

INSULT: See Assault.

INVEST: To begin a siege. This is usually done by sending a large detachment of horse to block all of the avenues into a Place, preventing relief until the main army and artillery arrives to form the siege.

LADDER: See Scaling Ladder.

LINE: The name of any works in the open field. See

the

the specific entries for individual types of lines, such as Circumvallation, Countervallation, and Counter-approach. Also see Plate N-VII.

LINE OF LEAST RESISTANCE: In a mine, the shortest distance between the chamber where the powder is located and an open surface. If a mine is constructed properly, this will be the distance vertically between the chamber and the ground above it. If the galleries are ill-made or not filled with earth, then the line of least resistance will be back through the galleries, and the mine will not serve its intended purpose. See page 213.

LODGEMENT: A work made in the breach or other part of the fortification by the besiegers in order to keep possession of that location. At the minimum, this is a parapet to cover the besiegers from the musket fire of the garrison.

LOOP-HOLE: A square or oblong hole in a fortification made to allow muskets and other small arms to be fired through it. See Plate N-IV.

LUNETTE: Also spelled Lunet. A work made to cover the re-entring angle between a bastion and a ravelin. The work is similar to, though smaller than, a ravelin and principally differs in its location. A lunette differs from a tenaillon in that a lunette covers only part of a face of the ravelin, whereas a tenaillon covers an entire face of the ravelin. The term lunette is also applied to a work made by the besieged beyond the ditch and opposite the place of arms. See Plate N-VI.

MAGAZINE: Any place where stores, arms, ammunition, or provisions are kept. A powder-magazine is a place where large quantities of gunpowder are stored. Such places must be arched and bomb-proof, and are frequently embedded in the ramparts. An artillery-magazine is a place made 25 to 30 yards behind a battery and at least 3 feet underground to hold all of the

powder,

powder, shells, and matches required for at least a day's supply for the battery. The magazine should be roofed with boards to prevent a cave-in, and the trenches leading to it should have traverses to prevent cannon balls fired à ricochet from entering the magazine.

MANTLET: A kind of moving parapet made of strong planks 4 to 5 feet wide and 3 to 5 feet high. The mantlet is mounted upon two wheels and has a long pole affixed to it. It is used by sappers to cover them from musket fire. A double mantlet is made with two sets of boards stoutly joined to form a saliant angle at the front. The mantlet may be covered with iron plates to improve its strength. See page 4 and Plate I.

MATCH: Usually this refers to slow-match, though it can also refer to quick-match. In general, match is a kind of rope treated to burn at a relatively steady rate. Slow-match is often made from hempen tow or other coarse fibers formed into slack yarns and then twisted into a cord. The cord is boiled in either saltpetre or the lees of old wine and then dried. Quick-match is made the same way, but upon taking the cord from the saltpetre solution, it is rolled in mealed powder moistened with alcohol and then dried.

MINE: A subterranean passage dug under the wall or rampart of a fortification, with the purpose of setting a charge of gunpowder to blow up the fortification. A counter-mine is one dug under the batteries, lodgments, or other works of the besiegers. The mine is dug a specific distance below ground so that the line of least resistance is known and the appropriate charge for the mine can be calculated. The mine is charged by placing a box of powder at its chamber or focus, then packing the area around the box and in the gallery full of soil, so that the mine will explode upwards and not back into the gallery. See the section on mining starting on page 206 for details.

MORTAR: A short piece of artillery designed to launch shells, bombs, or balls at a high angle so that they land within the walls and fortifications of the enemy. Mortars are measured based on the size shell they can fire. For instance, a 13-inch mortar fires a shell that is 13 inches in diameter using a powder charge of around 3 pounds. A mortar is shorter than a howitze and is usually used to launch shells at a steeper angle. Mortars are mounted on a sturdy wooden bed, which sometimes has axles to allow wheels to be affixed to transport it.

MUSKET: The standard firearm used by foot soldiers. Muskets are fired with a flintlock that strikes a piece of flint against a hardened steel hammer to ignite a small quantity of powder in a pan; this powder communicates with the powder in the barrel through a touch-hole, and then the main charge combusts, flinging the bullet or shot out of the muzzle. A musket differs from a rifle, in that the barrel of a musket is smooth, while that of a rifle has grooves that spin a bullet as it is fired. The smooth barrel allows a musket to be loaded more swiftly than a rifle.

NEEDLE, IRON: A long pointed metal bar used to probe the ground to determine if there are any mines. This is often used by miners to detect counter-mines. It differs from an augre in that it doesn't have a widened blade and is therefore harder to bore into the earth.

ORILLON: An extension of the face of a bastion to cover the flank. An orillon prevents grazing fire from reaching the cannons on the flank, and is often covered with or made out of stone. See Plate N-VI.

OUT-WORK: Any fortification outside of the main body of the Place, such as ravelins, horn-works, and crown-works.

PALISADE: Also spelled Palissade or Palissado. Stakes which are approximately 9 feet long and 6 or 7

inches

inches in diameter or square. They are placed in the ground to a depth of 3 feet, all in a row with a 2 ½ to 3 inch gap between them. The stakes are set about 3 feet from and parallel to the parapet or side of the glacis to secure the work from surprise. Palisades are also used to fortify the avenues of gorges, ravelins, and the bottoms of ditches. They are usually planted vertically, though some are sloped toward the enemy so that when a rope is cast over them in order to tear them down, the rope will be more likely to slip off. See Plate N-IV.

PARALLEL: Also called Places of Arms. The trenches or lines made parallel to the defences of the Place being besieged. The parallels are 15 to 18 feet wide and join the various attacks together so that they may mutually support each other. Guards are stationed in the parallels to come to the aid of the workmen when attacked. There are usually three parallels in the attack, the first about 1800 feet from the covert-way, the second about 960 feet from the covert-way, and the third directly on the glacis. Their invention is credited to M. *Vauban*. See Page 46.

PARAPET: Also called a Breastwork. An earthwork designed to cover soldiers or artillery from exposure to cannon and small arms. It is typically 6 feet in height on the inside, sloping to 4 or 5 feet on the outside, and has a width of 18 to 20 feet. There is often a raised step, or banquette, on the inside. There may be embrasures at various intervals to allow cannons to fire from behind the breastwork without unnecessarily exposing the artillery crews to enemy fire. See Plate N-IV and Plate N-V.

PARK OF ARTILLERY: See Artillery Park.

PARLEY: A conference with the enemy. Also, a signal beat on the drum for holding such a conference.

PARTIZAN: Also spelled Partisan. A kind of halberd or pole-arm with three tines, the center tine

being

being longer than the two edge tines. It can be used in place of a fork for setting or toppling gabions.

PETARD: A type of shaped charge for bursting open gates. It is generally made of a board 2 inches think and about 2 ½ feet square to which is affixed a large pot made of brass, gun metal, or soft iron, with the mouth of the pot against the boards. The pot holds from 9 to 20 pounds of powder, and is filled through a hole in the base on the opposite side of the planks. The hole is plugged and a vent or touch-hole is made either in the plug or along the side of the pot. The petard has a hook to hang it on the gate or is supported with three staves affixed to the board. It is placed against the gate, and then fired. See page 116, page 193, and Plate XX.

PICKET: A stake sharpened at one end, used for various purposes. Those used for placing fascines and gabions are 3 to 5 feet long. The pickets used for marking the locations to dig trenches are 3 feet long. Those used for marking out the artillery park are 5 to 6 feet long. See Plate I.

PIONEER: Pioneers are either private soldiers or men brought in from the country side who work on entrenchments, earthworks, mines, and approaches. Pioneers are equipped with hatchets, saws, spades, and pick-axes.

PLACE OF ARMS: In the attack, this is the same as a Parallel. In the covert-way, this is a wider area on the same level as the covert-way either at a salient angle or opposite a re-entring angle. The place of arms generally projects out a short distance in a saliant angle to allow troops there to fire along the glacis in either of two directions. See Plate N-V.

PLATFORM: A bed of wood used in a battery. The planks are generally 2 ½ inches thick and from 8 to 15 feet long, depending on the size of the gun to be supported. The planks rest on 4 to 6 square beams 4 to 7 inches

inches on a side called sleepers. Rather than using iron nails, wooden pegs are used to prevent accidental sparking between a cannon wheel and a nail. The platform is sloped with the low end near the parapet to limit the recoil of the gun and to make it easier to roll the gun into position. See page 38, Plate VII, and Plate VIII.

POLYGON: A convex figure comprised of more than four sides. This is the shape of the typical retrenchment and of many fortifications.

POSTERN: See Sally-port.

POWDER: Gunpowder. See the Notes on gunpowder on page 263.

RAMPART: The mass of earth that forms the main wall and protects the fortified place. The ramparts are rarely more than 18 feet high, excluding the parapet, and are made 60 to 70 feet thick. One side of the rampart slopes down to the principal level of the main Place; the other side slopes down steeply to the ditch. The top of the rampart should be broad enough to allow troops or artillery to be moved along it and to pass any artillery already stationed there. See Plate N-IV and Plate N-V.

RAVELIN: Also called a Demilune or a Half-moon. A detached work that covers a curtain. It is formed by two faces meeting in a salient angle. The ravelin prevents the flanks of the adjacent bastions from being attacked directly, and is defended by the faces of those same bastions. Ravelins are the most common of all outworks and are made slightly lower than the main ramparts and bastions, so that those works may defend the ravelins. See Plate N-V.

REDAN: Also spelled Reden and Redent. A projection in a work, consisting of two faces, whose parapet is continuous with the parapet of the work. Redans are sometimes used in a curtain instead of

ravelins.

ravelins. They are also commonly used in the line of circumvallation and along rivers that run through a fortified town. See Plate N-VII.

REDOUT: Also spelled Redoubt and Reduct. Most commonly, this refers to a square work without bastions that is located outside of the glacis, often within musket shot of the Place. It has loop-holes to fire through its parapet and is surrounded by a ditch. Redouts are sometimes made by the besiegers in the angles of the trenches to better protect the workers from sallies. A redout can also mean a small retrenchment within a ravelin. See Plate N-III.

RE-ENTRING ANGLE: Also spelled Rentring Angle and Re-entering Angle. An angle in the walls of the fortification that points inward toward the body of the Place. These occur between works, such as between a bastion and a ravelin, and also between the flank of a bastion and the curtain. See Plate N-V.

RETRENCHMENT: A fortification thrown up in a bastion or other work by the besieged once the besiegers have gained a lodgment in the walls of the work. The retrenchment often takes the form of the work it is within; for example, the retrenchment within a bastion often has two faces, two flanks, and a gorge. The retrenchment can also be made of just two faces or a single face. Retrenchments allow the ground to be contested inch-by-inch. Some retrenchments are made before the Place is besieged by putting double parapets with a ditch between them in a bastion. The term retrenchment can also be applied to any fortified work with a parapet and ditch. See Plate XXI and Plate N-VII.

REVETEMENT: A strong wall built on the outside of the rampart or parapet to support the earthworks. This allows the walls to be steeper than if they were made entirely of earth. A revetement makes the Place harder

to

to surprise, as troops will have a harder time scaling a revetement than an earthen wall. See Plate N-IV.

RICOCHET: Firing à ricochet is when a cannon, howitze, or mortar is loaded with a light charge and fired at a horizontal angle of 5 to 12 degrees, such that the ball or shell passes over the parapet and bounces or rolls along the rampart. The ball can do much greater execution than if it drops onto the enemy from a high angle, and the sight of a shell rolling toward them will make the stoutest troops break ranks. A battery erected for firing à ricochet is called a ricochet battery. Traverses are often made in trenches and along a parapet to prevent ricochet firing from taking its full effect. The invention of firing à ricochet is attributed alternately to M. *Vauban* and M. *Belidor*.

RIDGE OF THE GLACIS: The edge of the glacis near the fortification, which forms the parapet of the covert-way. See Plate N-IV and Plate N-V.

SALIENT ANGLE: Also spelled Saliant Angle. An angle in the walls of the fortification that points away from the body of the Place. These occur between the faces of the works, such as the point of a bastion or the point of a ravelin. See Plate N-V.

SALLY: An attack made by the garrison on the besiegers. Usually it is attempted to keep the sally a secret until the sallying-party falls upon the enemy's workmen or guards. The purpose is to drive out the workmen and destroy the works. See page 47.

SALLY-PORT: Also called a Postern or a Postern Gate. A small gate in the middle of a curtain whose purpose is to let out sallying parties and also to communicate with any outworks. See Plate N-IV and Plate N-V.

SAND-BAG: A cloth bag filled with sand or earth and used to repair breaches and rebuild parapets. They are 10 to 16 inches in diameter and 24 to 30 inches tall.

Sand

Sand-bags are also sometimes used to erect a parapet instead of digging a trench, but only when the ground is so hard that digging is impossible.

SAP: A trench or approach made under cover. Saps are often made 10 to 12 feet wide, with a parapet on one side. A double sap has parapets on both sides. In both cases, the sap has frequent traverses so as to impede ricochet firing. See page 43.

SAP FAGOT: A small fascine. A sap fagot is a bundle of branches or sticks used to support the earthworks in a sap. They are commonly 3 feet long and 6 to 8 inches in diameter.

SAP-FORK: A long-handled fork used to place and move gabions without exposing the workmen to danger. See Plate I.

SAP-HOOK: A long-handled tool with a cutting blade and a hook, used to place gabions and to cut away branches and other impediments without exposing the workmen to danger. See Plate I.

SAUCISSON: A long hose made of cloth or leather and filled with gunpowder. It is laid in an auget, with one end in the chamber of the mine and with the other end accessible at the entrance to the gallery, and is used to ignite the mine's charge. The saucisson is often 1 ½ inches in diameter. Another meaning of saucisson is a long fascine, or bundle of branches or sticks, used to support the earthworks in a battery or to repair a breach. They are commonly 10 to 16 feet long and 12 to 18 inches in diameter.

SCALING LADDER: Used in taking a Place by escalade. Ladders can either be made of rope or wood, and are often made to come apart in sections to be easier to carry. See page 123 and Plate XX.

SCARP: See Escarp.

SECOND COVERT-WAY: If a second ditch is made outside of the glacis, this is the covert-way that is on the far

far side of that ditch. See page 81 and Plate XIV.

SECOND DITCH: A ditch made outside of the glacis. The second ditch is typically only made if it can be filled with water to partially flood the glacis. See page 81 and Plate XIV.

SHELL: A hollow iron ball filled with gunpowder. A fuse is inserted into a plug that closes the shell, and is ignited by the gunpowder used in firing the shell from a cannon. The fuse is made such a length that the shell will either explode just before it hits the target, or, if the shell is fired à ricochet, then sometime after the shell will have rolled or bounced along the ground. The part of the iron shell opposite the fuse is made thicker than the rest, so that the shell will land with the fuse uppermost and not extinguish itself.

SHOT: Any kind of round ball used in ammunition. Shot for cannons consists of solid balls, grape-shot, and canister-shot, and is usually made of iron. Shot for muskets and wall-pieces is made of lead.

SHOULDER: The corner on a bastion or other work between the face and the flank. If the face is extended so that the shoulder covers the flank, it is called an orillon. The term shoulder can also refer to a demi-bastion, which is often made in the flanks of horn-works or crown-works. See Plate N-V and Plate N-VII.

SIEGE: To surround a fortified place with an army, cutting it off from the neighboring countryside, and then approach the Place by trenches and saps so as to be covered against artillery and musket fire from it.

SLEEPER: A supporting plank in a battery platform. The sleepers are usually square and 4 to 7 inches on a side. They are covered with boards, upon which the cannons can be rolled into position.

SMALL ARM: A musket or wall piece. Generally, a small arm is any firearm that can be carried by a single individual.

STOCKADE: A line of sharpened posts or stakes. Alternately, a fortification consisting solely of such posts and without ramparts. A stockade can be run across a shallow river to prevent the passage of boats, or across a marsh to prevent surprise.

STONE MORTAR: A type of large mortar made to fire a quantity of small stones rather than an iron shell or bomb.

STUFFED GABION: A gabion that has been filled with sticks and wood. It is rolled before the first worker in a trench or sap to protect against musket fire. See Gabion.

SUCCOR: An effort made to relieve the Place that is being besieged. This applies both to an attempt to bring supplies to the Place and to an attempt to drive off the besiegers.

SURPRISE: Also spelled Surprize. To attack an enemy unexpectedly. See page 118 and page 125.

SWORD-IN-HAND: To take a Place sword-in-hand is to make a direct assault on the Place without the benefit of covering fortifications. The commanding officer often leads his troops, waving them on with his sword.

TENAILLE: Also spelled Tenail. A low work made in the ditch in front of a curtain. It is often in line with the face of the neighboring bastion. Tenailles can be made either in a straight line, or as a face and a flank, or as a face, a flank, and a length of curtain. A second meaning of tenaille is a double-pointed or triple-pointed ravelin. In this case, the work is placed before a curtain, and has two flanks that are in line with the flanks of the bastions on either side of the curtain. The flanks are connected by two faces forming a re-entrant angle or by four faces forming two re-entrant and one saliant angle. The phrase, the tenaille of a Place, refers to everything between the points of two neighboring bastions, including the faces, flanks, and curtain. See Plate N-V,

Plate

Plate N-VI and Plate N-VII.

TENAILLON: A work made to cover the re-entring angle between a ravelin and a bastion. The work is similar to, though smaller than, a ravelin, and principally differs in its location. A tenaillon differs from a lunette in that the tenaillon covers an entire face of the ravelin, whereas a lunette covers only part of a face of the ravelin. See Plate N-VI.

T MINE: A mine with two or four chambers whose gallery resembles the shape of the letter T.

TOISE: A French unit of length equal to 6 French feet, slightly longer than the English fathom. See the Notes on units on page 254.

TOWER BASTION: Short towers made in the shape of bastions, with rooms underneath to place guns and men. The tower is often within a regular bastion, and serves as a form of retrenchment.

TRAVERSE: A ridge of earth made along the covert-way, along the parapets of the walls, and in the trenches. The ridge is usually as high as the parapet or the glacis. The base of the traverse is often 18 feet across. The purpose of the traverse is to prevent shells fired à ricochet from travelling the length of the work. See Plate N-IV and Plate N-V.

TRENCH: The trenches are ditches made by the besiegers to approach a Place while remaining relatively secure from the garrison's guns and muskets. The tail of the trench is where it began, and the head is where it is currently being dug. See page 31.

TRUNION: The cylindrical pieces that extend to either side of a cannon, howitze, or mortar barrel, and allow the barrel to be mounted on a carriage or bed.

TURFED: A turfed earthwork is one which is covered in grass and otherwise unsupported.

WALL-PIECE: A type of large musket that can be carried by a single man but must be placed on a rest in

order

order to fire it.

WOOLPACK: A cloth bag packed with wool and used for lodgments in places where there is little earth. They are made 5 feet long and 15 inches in diameter.

BIBLIOGRAPHY

IN addition to *Muller*'s own books, the following works were referenced while compiling these notes. Of *Muller*'s works, *The Elements of Fortification* was most useful in compiling the lexicon of terms.

AUSTIN, Jane G., "Highly Explosive," *Atlantic Monthly*, November 1870, pp. 527-542. This is an account of a tour of a gunpowder mill in 1870. It contains a good description of refining saltpetre so as to remove the other unwanted salts.

COMMITEE of Safety of the Colony of New-York, *Essays upon the Making of Salt-petre and Gun-powder.* New York: Samuel Loudon, 1776. Aside from *Diderot*'s Encyclopédie, this is perhaps the most valuable work for understanding 18th century gunpowder manufacturing. It contains a series of articles on making saltpetre which range from clear and lucid to obscure and somewhat alchemical. It also has a brief but complete description of gunpowder as manufactured in Ulster County, New York by an amateur producer.

DIDEROT and d'Alembert, *Encyclopédie, ou Dictionnaire Raisonné des Sciences, des Arts et des Metiers, par une Sociètè de Gens de Letters.* Paris: Briasson, David, Le Breton, and Durand, 1751. The Encyclopédie is an amazing compendium of knowledge, with a lengthy set of alphabetical entries on diverse topics and a huge collection of illustrated plates. A few of the entries consulted include those on charcoal, sulfur, saltpetre, gunpowder, and fortification.

DU CHESNOY, Captaine Michael, *Plan of Carillon ou Ticonderoga: which was quitted by the Americaines in the night from the 5th to the 6th of July* 1777. Washington, D.C.: Library of Congress Geography and Map Division, digital map collection, digital ID g3804t ar300200. This was the source image used for the map in Plate N-I and Plate N-III.

Gentleman's Magazine, 1784. An extremely terse obituary notice for *Muller* is given on page 475.

HILL, George Birkbeck, *Boswell's Life of Johnson, including Boswell's Journal of a tour to the Hebrides and Johnson's diary of a Journey into North Wales.* Oxford: Clarendon Press, 1887. A footnote on page 351 explains how *Muller* was consulted on a mathematical question regarding the form of a new bridge to be built in London. It isn't clear if *Johnson* consulted *Muller*, or if the two happened to be in agreement. *Muller* is mentioned in flattering terms and is used to bring credibility to *Johnson*'s opinion.

HOGG, Brigadier O. F. G., *The Royal Arsenal.* London: Oxford University Press, 1963. This contains a few brief details on John Muller.

MANNINGHAM, Henry, *A Complete Treatise of Mines: Extracted from the Memoirs d'Artillerie.* London: A. Millar, 1741. This is a striking contrast to *Muller. Manningham* uses the frustum of a cone (not a paraboloid) for his calculation on the figure of mines. He subscribes to the theory that the base of the mine always has a diameter of twice the line of least resistance despite citing some of the experiments that occurred at La Fere. Further, he gives a huge variety of values for the weight of a cubic pound of powder, namely 60 lb 2 oz, 65 lb, 70 lb, 71 lb, 80 lb, and 95 lb 4 oz. Of these, he favors 70 lb in one calculation and 80 lb in another. It is not clear whether he uses French or English units.

NYE, Nathaniel, *The Art of Gunnery.* London: William

William Leak, 1670. This lists multiple ways to make saltpetre, all of which are somewhat cruder than the methods outlined in Diderot and other 18th century works. It also gives ratios of saltpetre, sulfur, and charcoal for various grades of powder.

PARTINGTON, J. R., *A History of Greek Fire and Gunpowder.* Baltimore, Maryland: The John Hopkins University Press, 1999. This book was originally published in 1960, and, although a difficult read, contains extensive information on gunpowder prior to the 17th century. It includes gunpowder compositions used through 1900, with notes about saltpetre and charcoal.

ROBERTSON, John, *A Treatise of Mathematical Instruments.* London: J. Nourse, 1775. As reprinted under the Flower-de-Luce Books imprint by The Invisible College Press, Arlington, Virginia, 2002. This work contains many period mathematical examples, with unit conversions and densities of materials. *Robertson* gives 114 English Feet = 107 French Feet and 108 English Pounds = 100 French Pounds. He also lists the density of gunpowder as 69.12 pounds per cubic foot.

RUMFORD, Sir Benjamin Thompson, Count, *Collected Works of Count Rumford*, edited by Sanborn C. Brown. Cambridge, Massachusetts: The Belknap Press of Harvard University Press, 1970. Count *Rumford* has several papers on gunpowder, the most pertinent of which dates from 1787 and gives the specific gravity of Government powder in a loose heap as 0.836 (equal to 52.25 pounds per cubic foot) and the same powder well shaken together as 0.937 (58.56 pounds per cubic foot). Count *Rumford* also describes a variety of trials to measure the force of gunpowder and its effect on projectiles. Government powder is gunpowder that has passed the minimum proof requirement of the British government.

SMITH, Capt. George, *An Universal Military Dictionary.* Whitehall: J. Millan, 1779. As reprinted by Museum Restoration Services, Ottawa, Ontario, 1969. This is a large compendium of military terms. Some parts, such as that on the charges of mines, are copied directly from *Muller*'s works. It was invaluable in compiling the lexicon of terms. *Smith* gives 1.068 English Feet = 1 French Foot.

SMITH, George, *Dictionary of National Biography, founded in* 1882. London: Oxford University Press, Waverly Book Co., Ltd. This contains a short biographical entry on *Muller*, which is the backbone of the biography in these notes.

STONE, Edmund, *The Construction and Principal Uses of Mathematical Instruments.* London: J. Richardson, 1758. As reprinted by The Astragal Press, Mendham, New Jersey, 1995. This book is a compendium of tools and instruments. It includes numerous length conversion factors and has a short section on designing fortifications. *Stone* gives 12 English Inches = 11.25 French Inches.

STUBBE, Henry, *Legends no Histories, or, A Specimen of some Animadversions upon the history of the Royal Society.* London, 1670. This is a very aggressive attack on some historical works published by the Royal Society, including commentary on the history of saltpetre and gunpowder as published by Mr. *Henshaw*. As part of the commentary, *Stubbe* explains how saltpetre and gunpowder are made, which shows the process to be essentially identical to that of the 18th century, but the saltpetre refinement to be more variable.

SURREY RECORD SOCIETY, *Gunpowder Mills: Documents of the Seventeenth and Eighteenth Centuries.* Woking: Surrey Record Society, 2000. This contains inventories from two gunpowder mills, one from 1678

and

and one from 1753, a deed of sale from 1661, and the letter book from the owner of a gunpowder mill from 1790-1, together with a nicely written introduction. This has many fascinating details of the gunpowder manufacturing process.

U. S. GEOLOGICAL SURVEY, *Ticonderoga 15′ Quadrangle.* Denver: USGS, 1950. The map in Plate N-II was taken from the southwest and southeast corners of this quadrangle.

VAUBAN, Sebastien LePrestre de, *A Manual of Siegecraft and Fortification*, translated by George A. Rothrock. Ann Arbor: University of Michigan Press, 1968. This is a translation of a 1740 printing of *Vauban*'s memoir. The memoir was written between 1667 and 1672, but not formally published until 1740. Although *Muller* draws extensively from *Vauban*, much had been learned in the decades since the memoir was written.

VAUBAN, Sebastien LePrestre de, *The New Method of Fortification.* London: S. and E. Ballard, C. Hitch, and J. Wood, 1748. This is the fifth edition of an English translation of *Vauban*'s memoir on fortification, first translated in 1691. Along with information on how to design fortifications, it contains a dictionary of fortification-related terms.

VON MALITZ, Ian, *Black Powder Manufacturing, Testing & Optimizing.* Digmans Ferry, Pennsylvania: American Firework News, 2003. This book covers small-scale black powder manufacturing for use in pyrotechnics (fireworks). It explains various milling, corning, and glazing techniques, and gives fairly thorough safety warnings. It also contains a good description of the Goex powder mill as it stood in the 1990s.

WEST, Jenny, *Gunpowder, Government, and War in the Mid-Eighteenth Century.* Woodbridge, Suffolk: The Boydell Press, A Royal Historical Society publication, 1991.

1991. This is an excellent scholarly account of the gunpowder trade in Britain. It contains many small details of powder manufacturing, consumption, cost, storage, and legislation.

Acknowledgements

AS always, I am deeply indebted to R.R.M. for her indefatigable help and encouragement. Without her suggestions, the notes would be much for the worse. I also thank all of my associates and friends who listened to me go on at length about sieges, gunpowder, mines, and other topics. Any mistakes or eccentricities are purely my own.

— David W. Manthey, 2004

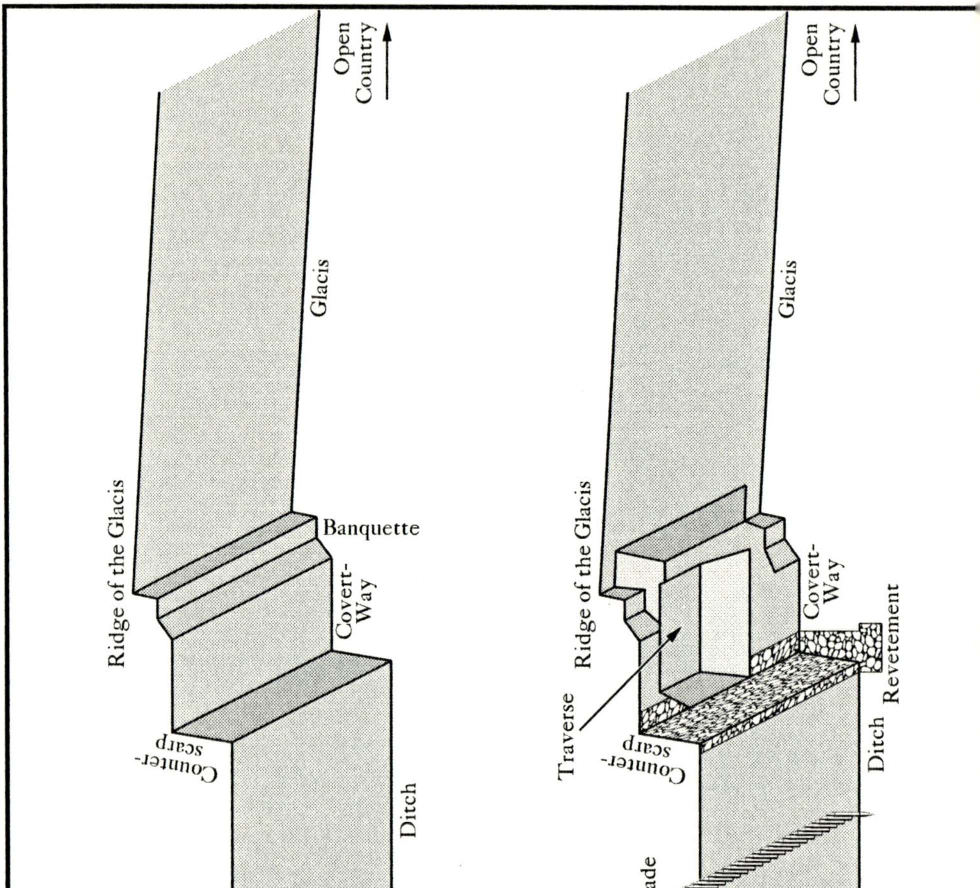

Fortification Terms – Cross-

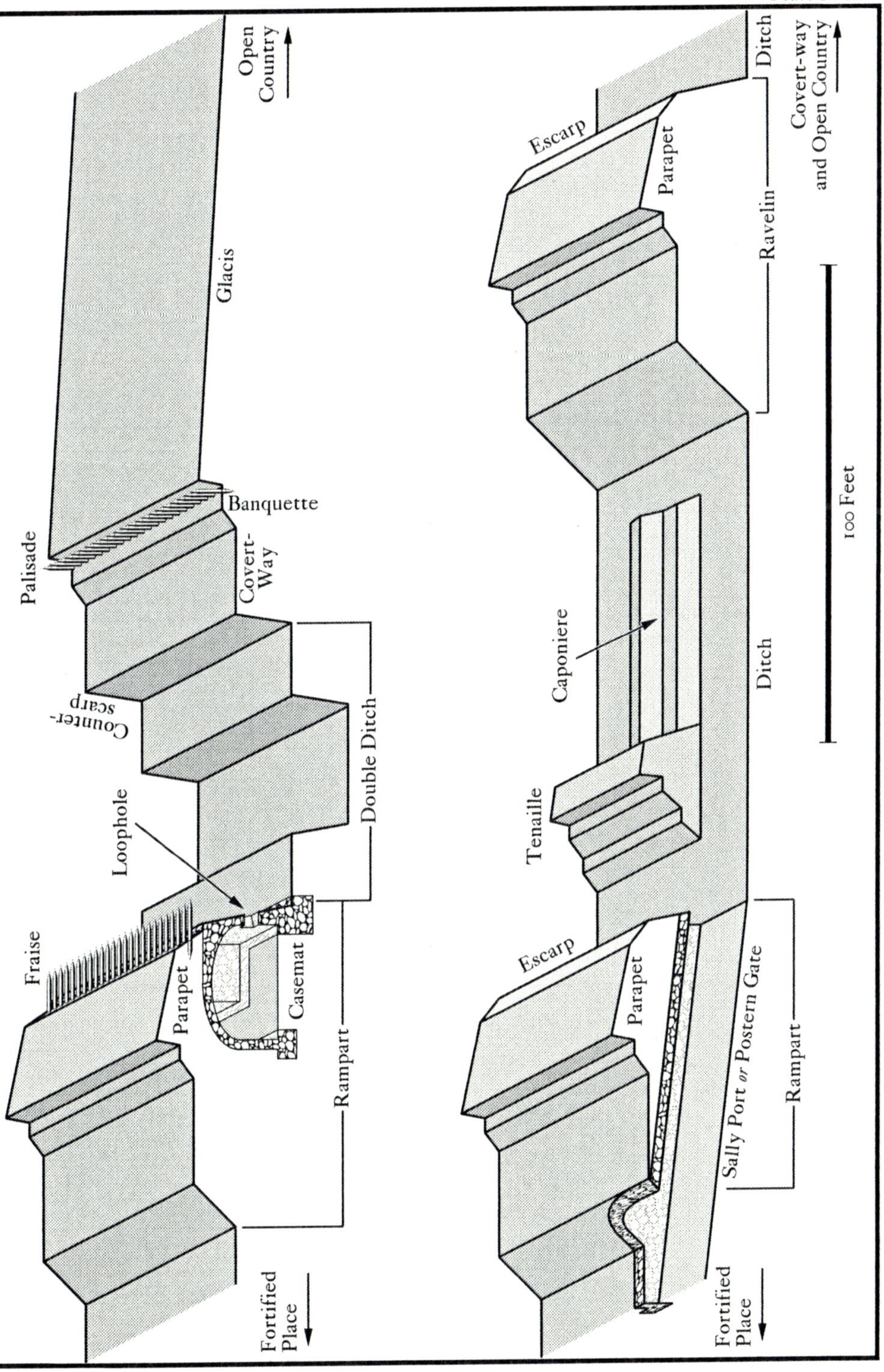

Sections of Different Fortifications

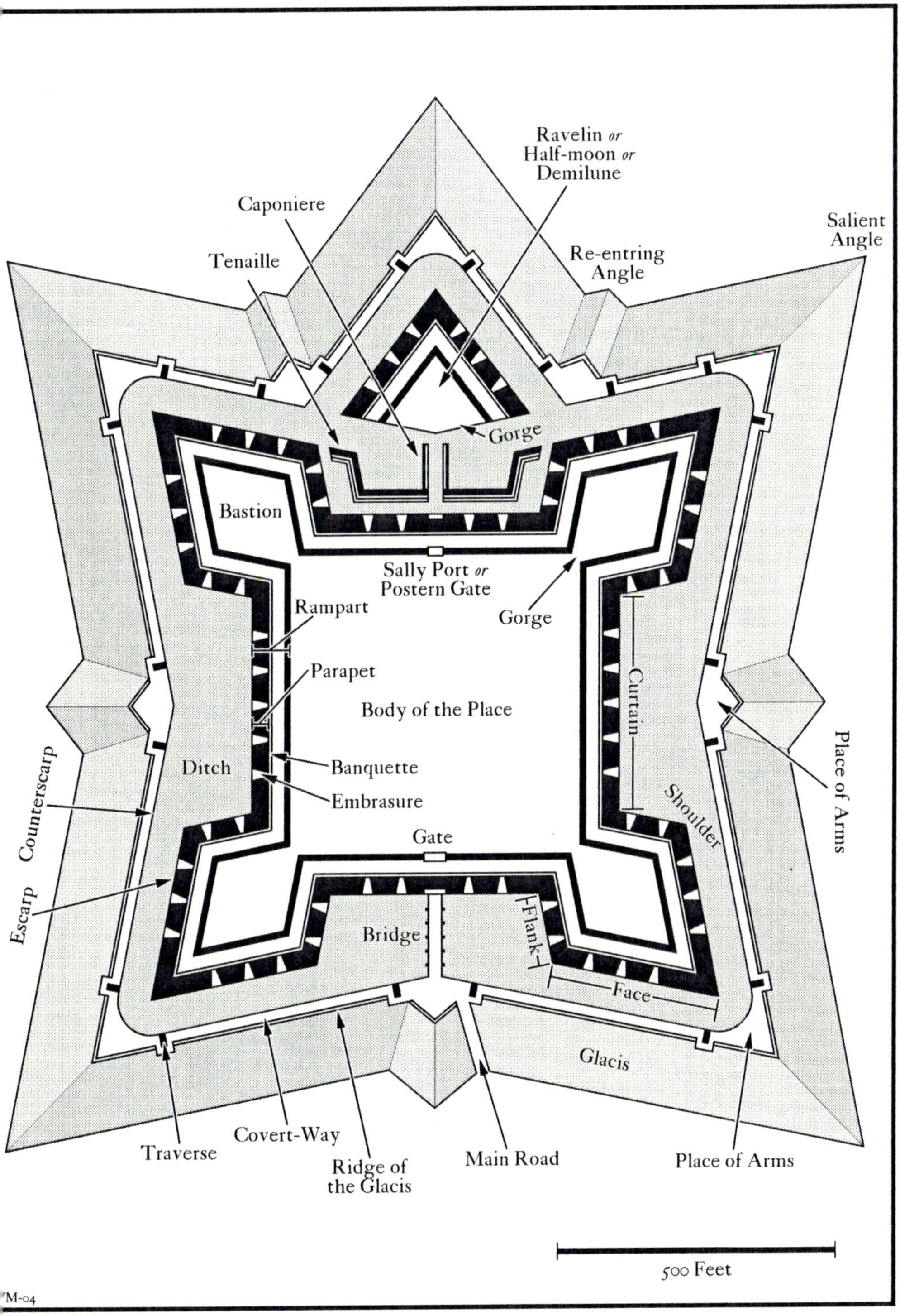

Fortification Terms – Plan of a Four-Bastioned Fort

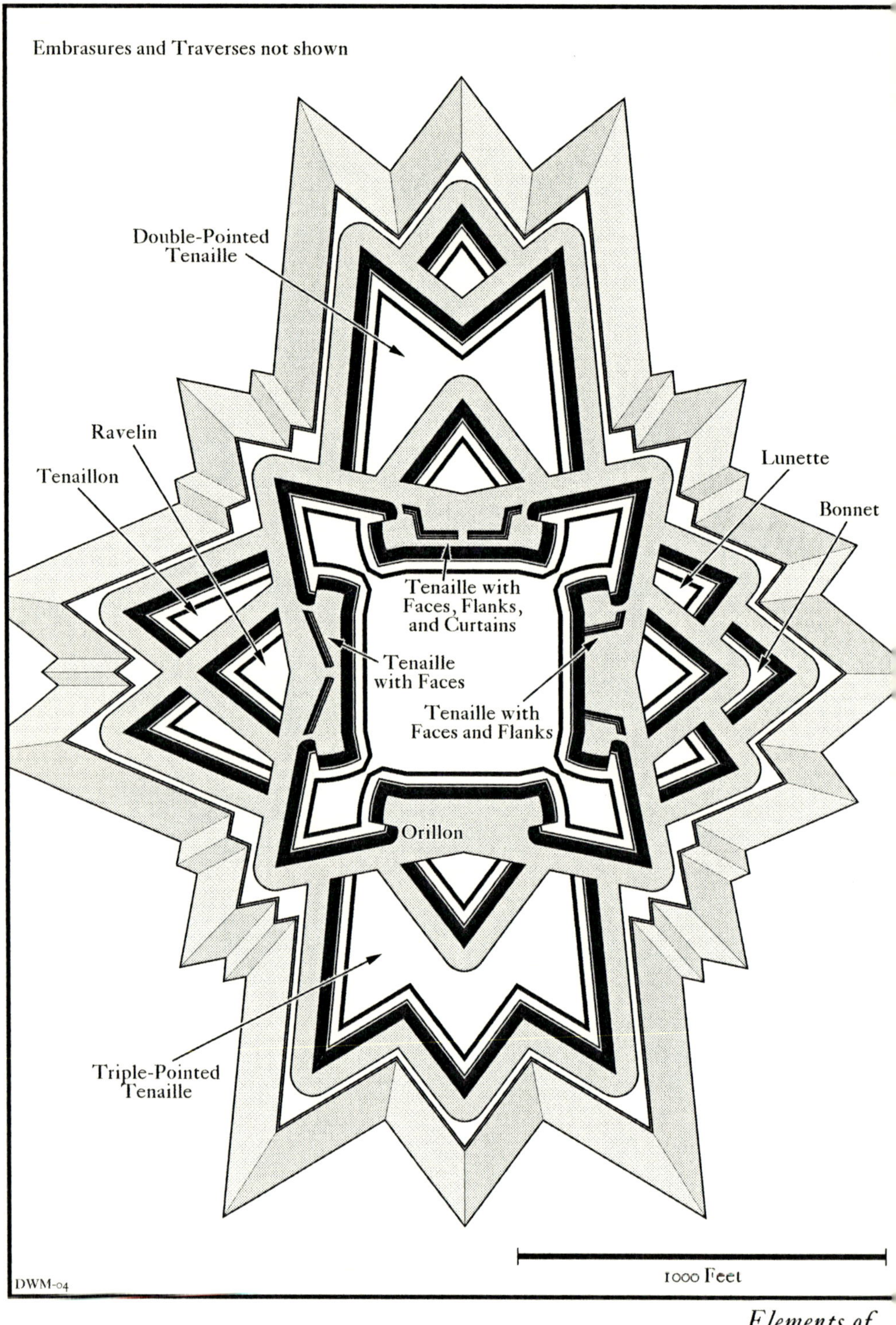

Embrasures and Traverses not shown
Double-Pointed Tenaille
Ravelin
Tenaillon
Lunette
Bonnet
Tenaille with Faces, Flanks, and Curtains
Tenaille with Faces
Tenaille with Faces and Flanks
Orillon
Triple-Pointed Tenaille
1000 Feet
DWM-04

Elements of

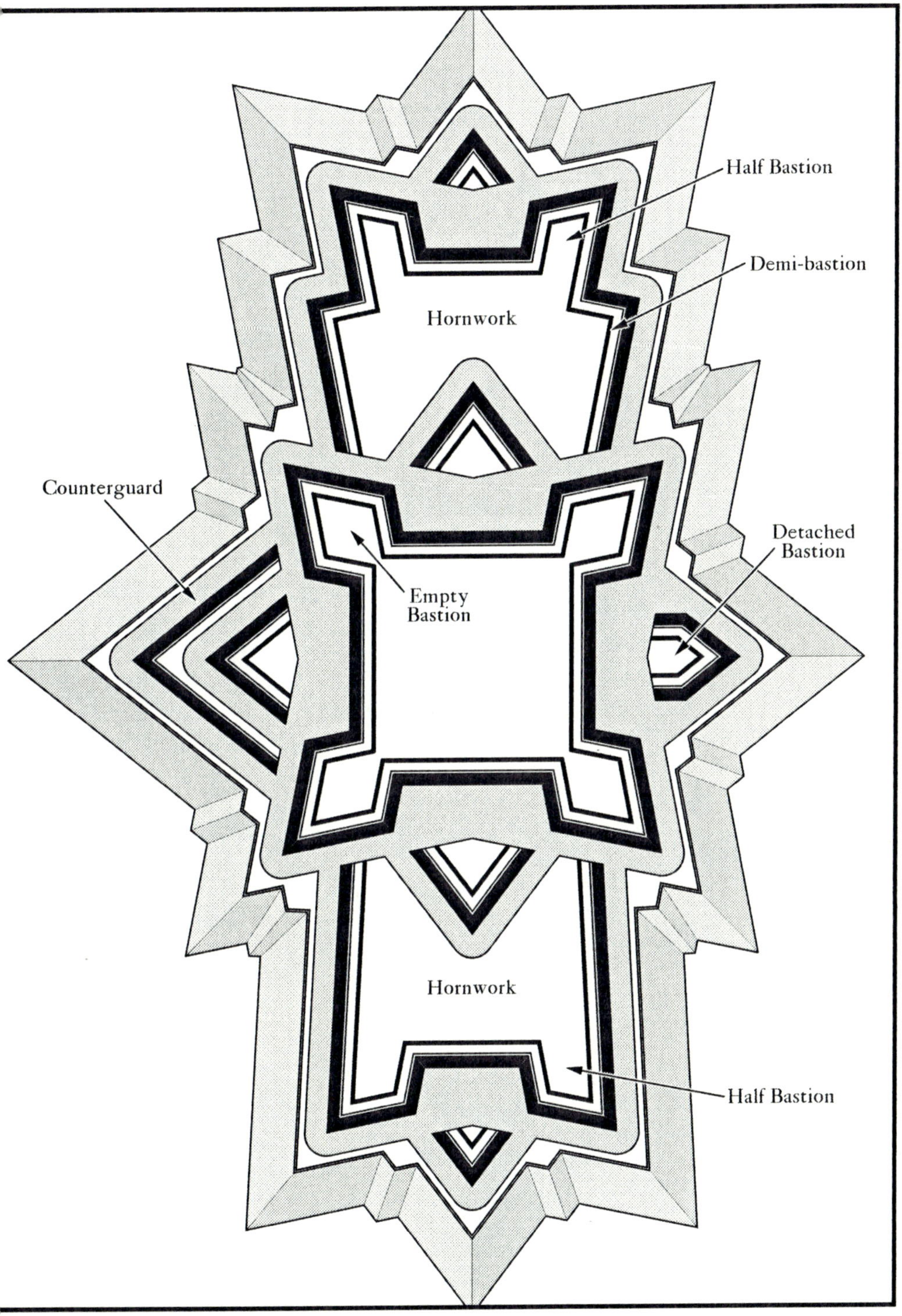

Fortifications

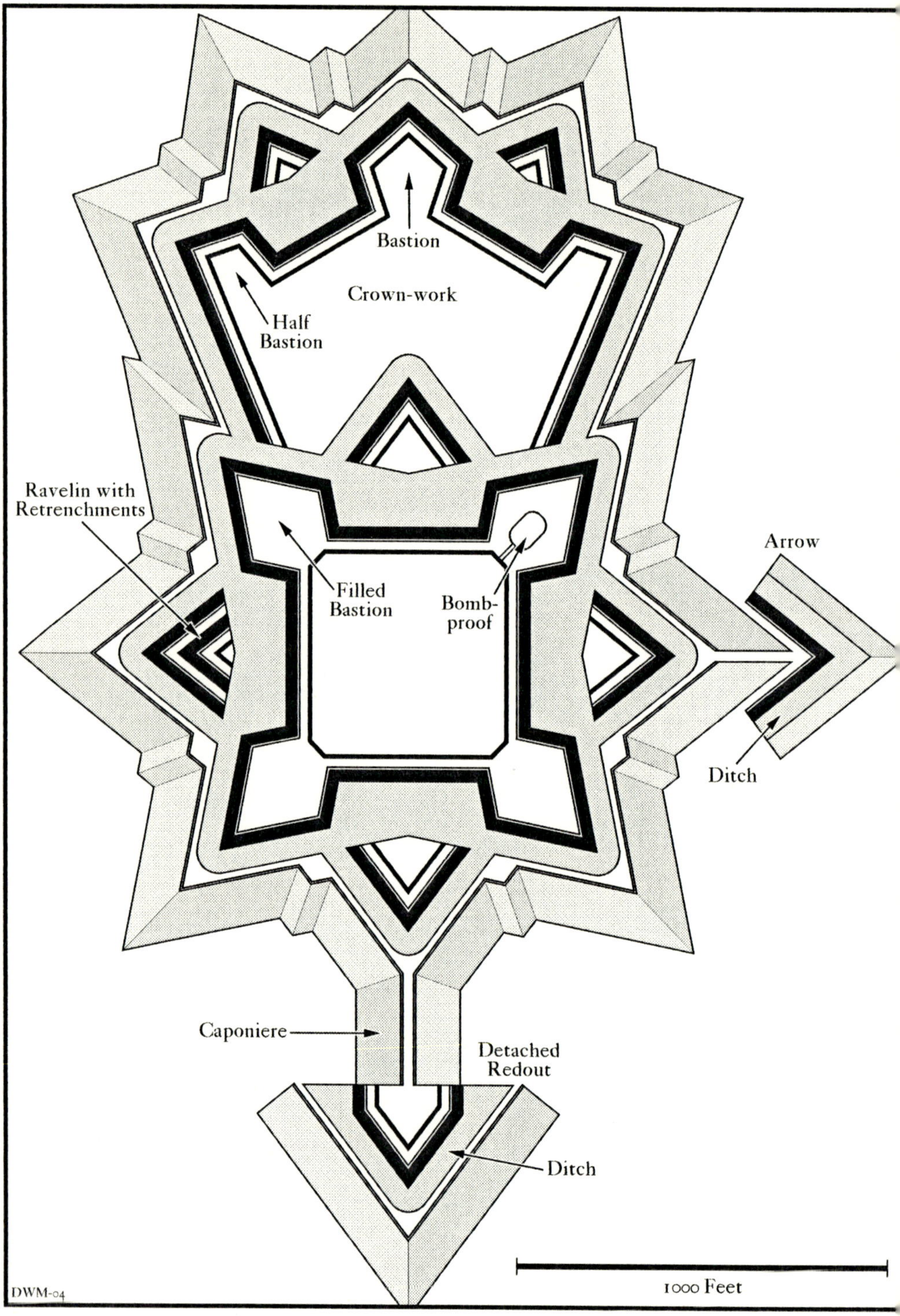
Bastion
Crown-work
Half
Bastion
Ravelin with
Retrenchments
Arrow
Filled
Bastion
Bomb-
proof
Ditch
Caponiere
Detached
Redout
Ditch
1000 Feet
DWM-04

Elements of

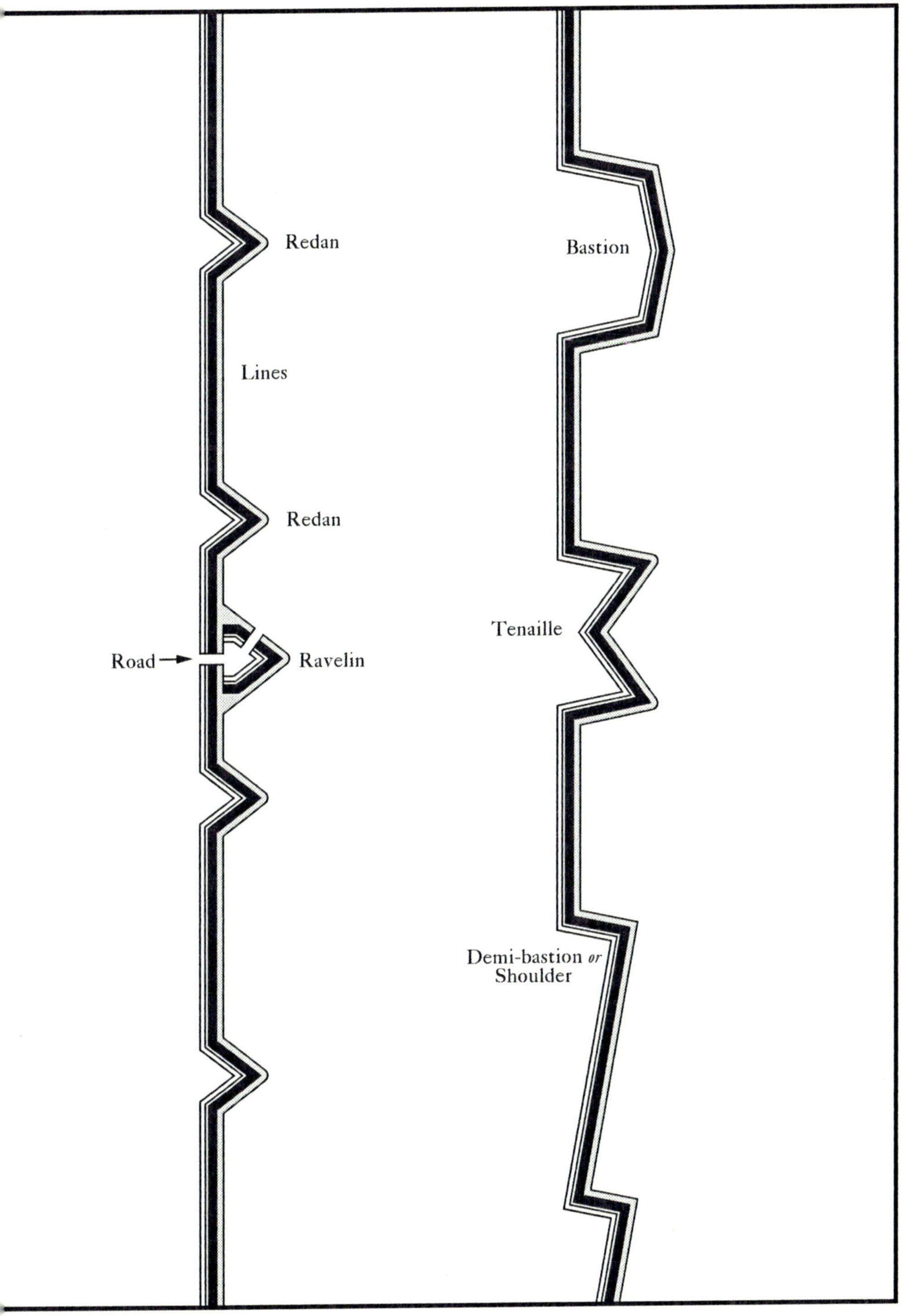

Fortifications

Printed in the United States
18233LVS00004B/319-336